John R. Ball and Charles H. Evans, Jr., *Editors*

Committee on Creating a Vision for Space Medicine During Travel Beyond Earth Orbit

Board on Health Sciences Policy

INSTITUTE OF MEDICINE

NATIONAL ACADEMY PRESS
Washington, DC

NATIONAL ACADEMY PRESS • 2101 Constitution Avenue, N.W. • Washington, DC 20418

NOTICE: The project that is the subject of this report was approved by the Governing Board of the National Research Council, whose members are drawn from the councils of the National Academy of Sciences, the National Academy of Engineering, and the Institute of Medicine. The members of the committee responsible for the report were chosen for their special competences and with regard for appropriate balance.

Support for this project was provided by the National Aeronautics and Space Administration. The views presented in this report are those of the Institute of Medicine Committee on Creating a Vision for Space Medicine During Travel Beyond Earth Orbit and are not necessarily those of the funding agency.

Library of Congress Cataloging-in-Publication Data

Institute of Medicine (U.S.). Committee on Creating a Vision for Space Medicine during Travel Beyond Earth Orbit.
Safe passage : astronaut care for exploration missions / John R. Ball and Charles H. Evans, Jr., editors ; Committee on Creating a Vision for Space Medicine during Travel Beyond Earth Orbit, Board on Health Sciences Policy, Institute of Medicine.
p. ; cm.
Includes bibliographical references.
ISBN 0-309-07585-8 (pbk.)
1. Astronauts—Health and hygiene. 2. Space medicine.
[DNLM: 1. Aerospace Medicine—standards. 2. Health Services Needs and Demand. 3. Astronauts. 4. Radiation Effects. 5. Space Flight. 6. Weightlessness—adverse effects. WD 751 I59s 2001] I. Ball, John, 1944- II. Evans, Charles H. (Charles Hawes), 1940- III. Title.
RC1135 .I576 2001
616.9'80214—dc21

2001005443

Additional copies of this report are available for sale from the National Academy Press, 2101 Constitution Avenue, N.W., Box 285, Washington, DC 20055. Call (800) 624-6242 or (202) 334-3313 (in the Washington metropolitan area), or visit the NAP's home page at **www.nap.edu**. The full text of this report is available at **www.nap.edu/readingroom.**

For more information about the Institute of Medicine, visit the IOM home page at **www.iom.edu.**

Printed in the United States of America

The serpent has been a symbol of long life, healing, and knowledge among almost all cultures and religions since the beginning of recorded history. The serpent adopted as a logotype by the Institute of Medicine is a relief carving from ancient Greece, now held by the Staatliche Museen in Berlin.

"Knowing is not enough; we must apply.
Willing is not enough; we must do."

—Goethe

INSTITUTE OF MEDICINE

Shaping the Future for Health

THE NATIONAL ACADEMIES

National Academy of Sciences
National Academy of Engineering
Institute of Medicine
National Research Council

The **National Academy of Sciences** is a private, nonprofit, self-perpetuating society of distinguished scholars engaged in scientific and engineering research, dedicated to the furtherance of science and technology and to their use for the general welfare. Upon the authority of the charter granted to it by the Congress in 1863, the Academy has a mandate that requires it to advise the federal government on scientific and technical matters. Dr. Bruce M. Alberts is president of the National Academy of Sciences.

The **National Academy of Engineering** was established in 1964, under the charter of the National Academy of Sciences, as a parallel organization of outstanding engineers. It is autonomous in its administration and in the selection of its members, sharing with the National Academy of Sciences the responsibility for advising the federal government. The National Academy of Engineering also sponsors engineering programs aimed at meeting national needs, encourages education and research, and recognizes the superior achievements of engineers. Dr. Wm. A. Wulf is president of the National Academy of Engineering.

The **Institute of Medicine** was established in 1970 by the National Academy of Sciences to secure the services of eminent members of appropriate professions in the examination of policy matters pertaining to the health of the public. The Institute acts under the responsibility given to the National Academy of Sciences by its congressional charter to be an adviser to the federal government and, upon its own initiative, to identify issues of medical care, research, and education. Dr. Kenneth I. Shine is president of the Institute of Medicine.

The **National Research Council** was organized by the National Academy of Sciences in 1916 to associate the broad community of science and technology with the Academy's purposes of furthering knowledge and advising the federal government. Functioning in accordance with general policies determined by the Academy, the Council has become the principal operating agency of both the National Academy of Sciences and the National Academy of Engineering in providing services to the government, the public, and the scientific and engineering communities. The Council is administered jointly by both Academies and the Institute of Medicine. Dr. Bruce M. Alberts and Dr. Wm. A. Wulf are chairman and vice chairman, respectively, of the National Research Council.

COMMITTEE ON CREATING A VISION FOR SPACE MEDICINE DURING TRAVEL BEYOND EARTH ORBIT

JOHN R. BALL (Chair), Executive Vice President Emeritus, American College of Physicians, Havre de Grace, MD

JOSEPH V. BRADY, Director, Behavioral Biology Research Center, The Johns Hopkins University, Baltimore, MD

BRUCE M. COULL, Head, Department of Neurology, University of Arizona College of Medicine, Tucson, AZ

N. LYNN GERBER, Chief, Department of Rehabilitation Medicine, National Institutes of Health, Bethesda, MD

BERNARD A. HARRIS, President, The Harris Foundation, Houston, TX

CHRISTOPH R. KAUFMANN, Colonel MC (USA), Division of Surgery for Trauma, Uniformed Services University of the Health Sciences, Bethesda, MD

JAY M. McDONALD, Professor and Chair, Department of Pathology, University of Alabama at Birmingham, Birmingham, AL

RONALD D. MILLER, Professor and Chair of Anesthesia and Perioperative Care, University of California at San Francisco, San Francisco, CA

ELIZABETH G. NABEL, Scientific Director, National Heart, Lung, and Blood Institute, National Institutes of Health, Bethesda, MD

TOM S. NEUMAN, Professor of Medicine and Surgery and Associate Director, Department of Emergency Medicine, University of California at San Diego, San Diego, CA

DOUGLAS H. POWELL, Senior Partner, Powell and Wagner Associates, Cambridge, MA, and Clinical Instructor in Psychology, Harvard Medical School, Boston, MA

WALTER M. ROBINSON, Assistant Professor of Pediatrics and Medical Ethics, Harvard Medical School, Boston, MA

CAROL SCOTT-CONNER, Professor and Head, Department of Surgery, University of Iowa College of Medicine, Iowa City, IA

JUDITH E. TINTINALLI, Professor and Chair, Department of Emergency Medicine, University of North Carolina at Chapel Hill, Chapel Hill, NC

Committee Liaisons:

Institute of Medicine Board on Health Sciences Policy

GLORIA E. SARTO, Professor of Obstetrics and Gynecology, National Center of Excellence in Women's Health, University of Wisconsin, Madison, WI

Institute of Medicine Board on Neuroscience and Behavioral Health

STEVEN M. MIRIN, Medical Director, American Psychiatric Association, Washington, DC

National Research Council Space Studies Board

MARY JANE OSBORN, Professor and Head, Department of Microbiology, University of Connecticut, Farmington, CT

Institute of Medicine Staff

CHARLES H. EVANS, JR., Study Director and Institute Senior Adviser for Biomedical and Clinical Research
MELVIN H. WORTH, JR., Scholar-in-Residence
JUDITH RENSBERGER, Senior Program Officer
VERONICA SCHREIBER, Research Assistant
SETH M. KELLY, Project Assistant to August 2000
TANYA M. LEE, Project Assistant from August 2000
GREG T. SHERR, National Research Council Intern (Summer 2000)

Institute of Medicine Board on Health Sciences Policy Staff

ANDREW POPE, Director
CARLOS GABRIEL, Financial Associate
ALDEN CHANG, Administrative Assistant

Institute of Medicine Auxiliary Staff

MICHAEL EDINGTON, Managing Editor to March 2001
PAIGE BALDWIN, Managing Editor from March 2001
SUSAN FOURT, Librarian to December 2000
WILLIAM McCLOUD, Librarian from December 2000

Copy Editor

MICHAEL K. HAYES

Reviewers

The report was reviewed by individuals chosen for their diverse perspectives and technical expertise in accordance with procedures approved by the National Research Council's Report Review Committee. The purpose of this independent review is to provide candid and critical comments to assist the authors and the Institute of Medicine in making the published report as sound as possible and to ensure that the report meets institutional standards for objectivity, evidence, and responsiveness to the study charge. The content of the review comments and the draft manuscript remain confidential to protect the integrity of the deliberative process. The committee wishes to thank the following individuals for their participation in the report review process:

JAMES P. BAGIAN, Director, National Center for Patient Safety, Veterans Administration, Arlington, VA
SAMUEL BRODER, Executive Vice President, Celera Genomics, Rockville, MD
EDNA FIEDLER, Manager, Federal Aviation Administration, Civil Aeromedical Institute, Federal Aviation Administration, Oklahoma City, OK
JOHN C. FLETCHER, Professor Emeritus of Biomedical Ethics, University of Virginia School of Medicine, Keswick, VA

LAZAR J. GREENFIELD, Frederick A. Coller Distinguished Professor and Chair, Department of Surgery, University of Michigan, Ann Arbor, MI
JOHN P. KAMPINE, Professor and Chairman, Department of Anesthesiology, Medical College of Wisconsin, Milwaukee, WI
NICK KANAS, Professor, Psychiatry, University of California San Francisco, San Francisco, CA
STEPHEN I. KATZ, Director, National Institute of Arthritis and Musculoskeletal and Skin Diseases, National Institutes of Health, Bethesda, MD
RUSSELL B. RAYMAN, Executive Director, Aerospace Medical Association, Alexandria, VA
JAMES L. REINERTSEN, Chief Executive Officer, Care Group, Boston, MA
DANNY A. RILEY, Professor, Department of Cell Biology, Neurobiology, and Anatomy, Medical College of Wisconsin, Milwaukee, WI
RICHARD H. TRULY, Director, National Renewable Energy Laboratory, U.S. Department of Energy, Golden, CO
DEBORAH J. WEAR-FINKLE, Forensic Psychiatrist/Flight Surgeon, United States Navy, Pensacola, FL

Although the individuals listed above have provided many constructive comments and suggestions, they were not asked to endorse the conclusions or recommendations nor did they see the final draft of the report before its release. The review of this report was overseen by Robert A. Frosch, Harvard University, and Robert M. Epstein, University of Virginia, Charlottesville, VA. Appointed by the National Research Council and the Institute of Medicine they were responsible for making certain that an independent examination of this report was carried out in accordance with institutional procedures and that all review comments are carefully considered. Responsibility for the final content of this report rests entirely with the authoring committee and the institution.

Foreword

Many American astronauts have participated in space shuttle missions which kept them in space for periods of 1-2 weeks. A few American astronauts participated in missions on the Russian space station *Mir* during which they spent substantially longer periods of time in space, missions measured in months rather than weeks. Some Russian cosmonauts spent substantially longer periods of time in space. For all of these missions, there was the potential to bring an acutely ill participant back to earth in a relatively short period of time.

Exploration missions into deep space, such as a journey to Mars, raise a series of new questions about the health of human participants. Some of the physiologic effects of shorter periods in space such as loss of bone calcium are likely to continue indefinitely during longer missions. Some risks such as radiation exposure may be increased as humans proceed further into space. Psychological and mental health issues will grow increasingly important during longer missions in a confined space, often involving individuals of very different social and cultural backgrounds. For prolonged missions, it will not be feasible to return an acutely ill individual to earth in a timely manner.

The Institute of Medicine convened a committee of experts to examine the issues surrounding astronaut health and safety for long duration space missions. At the request of the National Aeronautics and Space Administration (NASA), the committee was asked to make recommendations regarding the infrastructure for a health system in space, the principles that should guide such a system to provide an appropriate standard of care for astronauts, and to identify the nature of the clinical and health services research required before and during such long-term missions for astronauts.

The committee was faced with an extraordinary spectrum of issues related to prolonged travel into deep space. It focused its efforts upon the available data with regard to astronaut health and upon the areas where the evidence base was deficient. It particularly focused on areas that were amenable to investigation and development in preparation for human exploration of deep space.

The committee was selected with the intention of including some individuals with significant direct experience with space travel, but it included other experts in health care delivery and clinical research who could bring some fresh perspectives to the subject. We are deeply indebted to this volunteer committee expertly chaired by Dr. John Ball for their dedicated commitment to this project. We believe that the ideas, conclusions, and recommendations contained in this report should be helpful to NASA, as it prepares for further space exploration, as well as being of use to others who are concerned about the care of individuals in isolated locations on earth.

Kenneth I. Shine, M.D.
President, Institute of Medicine
July 2001

Preface

The National Academies, particularly the Space Studies Board of the National Research Council (NRC), have contributed substantial information on a full range of issues focused on basic aerospace research that has been a principal source of scientific advice to NASA. For example, the Space Studies Board recently provided guidance on a biomedical research strategy (SSB and NRC, 1998a, 2000). The present study by the Committee on Creating a Vision for Space Medicine During Travel Beyond Earth Orbit is the first time that the National Academies has addressed clinical issues through an analysis of the clinical research and health care practice evidence base for space medicine. In presenting its request to the Institute of Medicine (IOM) for the study presented in this report, NASA leadership noted that efforts to develop a more capable medical care delivery system in space have been internal to NASA itself, as documented in Chapter 1 of this report. The request used the language "create a vision" for health care for astronauts traveling on long-duration missions beyond Earth orbit. What was intended by the request was that a perspective beyond that internal to NASA and different from that provided by NRC be applied to the issues.

Those issues are reflected in the charge to the committee, which were (1) to assess what is known about the effects of space travel on health (which is the focus of Chapter 2 of this report), what is not known, and what might be done about the latter; (2) to suggest how health care during space travel might be approached (which is the focus of Chapters 3 and 4); and (3) to suggest how a system of health care for astronauts might be organized (which is the subject of Chapter 7 and, to some extent, of Chapters 3 and 4). The committee has sought to bring to its task the fresh perspective that NASA requested. That perspective has, for example, identified at least two specific areas on which we believe NASA's focus has been less than that which is now necessary. The first area involves a set of issues that are behavioral and cultural, including crew selection and training (Chapter 5). The second area is the collection of clinical data on astronauts, for which a new ethical approach is needed (Chapter 6).

What is the committee's perspective, then? Although the committee members come from diverse professional backgrounds, each with specialized knowledge, the diversity is within a relatively narrow range relative to the diversity of knowledge held by the general population or even the scientific community as a whole. Only a quarter of the committee members have had any previous experience with the space program, and only one has flown in space. The committee members do, however, have similar backgrounds, in that almost all are physicians (and those who are not are involved in clinical fields) and all have had experience in academic medicine. The committee thus brings a common perspective informed by the shared individual backgrounds of its members and the shared experiences of each of the members of the committee as it reviewed the evidence relative to its task. That perspective is largely, although not absolutely, that of the academic clinician and clinical researcher.

How these clinical and academic perspectives play out in the chapters of this report is illustrated by the following examples. In Chapter 2, the committee uses a biological systems approach to examine the data on what is known about clinical information, a systems approach that is based on an understanding of fundamental processes. In addition, the committee's approach reflects the academic perspective, one that values openness of dialogue and processes of peer review, particularly of research proposals. The committee's perspective is thus reflected in its conclusions and recommendations in Chapter 2 that relate to NASA's research process. In Chapter 3, the committee uses the perspective of the clinician, using the principle of continuity of care in its conclusions and recommendations.

Although Chapter 7, on planning an infrastructure for a system of astronaut health care, was prepared from the committee's perspective of academic clinician and clinical researcher, it differs from the other chapters in one important way. Whereas for the other chapters the committee reviewed hundreds of pages of NASA policies and procedures, cast a practiced eye on the evidence, analyzed that evidence, and reached conclusions that led to recommendations supported by the evidence in the individual chapters, for Chapter 7 the committee focused on principles drawn from the evidence examined throughout the study to support the conclusions and recommendations that the committee advances in Chapter 7. Thus, the committee suggests from its particular perspective what principles should be taken into account in the creation and evolution of a health care system for astronauts who will travel on long-duration missions beyond Earth orbit.

None of the committee members is a practicing engineer, and none is a physicist. Thus, we do not believe that our conclusions and recommendations should replace those of others. We hope, instead, that this report will add to the richness of NASA's approach to solving some of its most vexing issues.

As with all IOM reports, this one was a collaboration of the committee and IOM staff. The committee especially thanks Charles Evans for his excellent management of the project and leadership of the staff and for the staff support of Mel Worth, Judy Rensberger, Veronica Schreiber, Tanya Lee, and Seth Kelly. NASA staff, both at NASA headquarters and at the Johnson Space Center, gave generously of their time and energy; without the information that they provided, this report would not have been possible. Chapters of the report were developed by working groups of the committee headed by Lynn Gerber, Carol Scott-Conner, Doug Powell, and Walter Robinson; their dedication to the project was substantial. Lastly, the chair wishes personally to thank the committee members for the collegial manner in which they approached the task and for their responsiveness to the demands that such a complex project requires.

John R. Ball
Chair
June 2001

Acronyms

ADA	American Dental Association
AHRQ	Agency for Healthcare Research and Quality
ANARE	Australian National Antarctic Research Expeditions
ART	atraumatic restorative treatment
CAD-CAM	computer-aided design-computer-aided manufacturing
DCS	decompression syndrome
EVA	extravehicular activity
FAA	Federal Aviation Administration
IOM	Institute of Medicine
ISS	International Space Station
JSC	Johnson Space Center
MeV	million electron volts
NASA	National Aeronautics and Space Administration
NIH	National Institutes of Health
NRC	National Research Council
NSBRI	National Space Biomedical Research Institute

PTH	parathyroid hormone
SMAC	spacecraft maximum allowable concentrations
SMS	space motion sickness
SPEs	solar particle events
STS	Space Transportation System
T3	tri-iodothyronine

Contents

EXECUTIVE SUMMARY 1
Task of the Committee, 2
Risk, 3
Health Care, 6
Research, 8
Behavioral Health, 11
Data Collection and Access, 13
Engineering and Biology, 17
Organization, 19

1 ASTRONAUT HEALTH BEYOND EARTH ORBIT 23
Background, 23
Health Risks of Space Travel, 26
Charge to the Committee, 29
The Committee's Study of Health Care for Astronauts Traveling Beyond Earth Orbit, 30

2 RISKS TO ASTRONAUT HEALTH DURING SPACE TRAVEL 37
Overview, 37
Countermeasure to Solve Physiological Adaptations to Space, 40

Musculoskeletal System, 42
Loss of Bone Mineral Density in Microgravity, 42
Reversibility, Genetic Variability, and Mechanism of Bone Mineral Density Loss, 44
Clinical Research Opportunities in Astronaut Physiology and Health, 46
Effects of Microgravity on Skeletal Muscle, 47
Circulatory and Pulmonary Systems, 48
Orthostatic Hypotension, 48
Other Effects on the Cardiovascular System, 49
Effects of Microgravity on the Pulmonary System, 51
Alimentary System, 51
Nutrition in Space, 51
Space Motion Sickness, 51
Nervous System, 52
Neurovestibular Function, 53
Sleep and Circadian Rhythm, 53
Eye-Hand Coordination and Sustained Gross Motor Activity, 55
Peripheral Nervous System, 55
Reproductive System, 56
Effects of Radiation on Gametes, 56
Human Reproductive Physiology in Space, 57
Sex Differences, 59
Urinary System, 60
Physiological Monitoring, 61
Monitoring During Space Travel: Development of Technology, 61
A Strategy for a Space Medicine Clinical Research Program, 62
A Clinical Research Program for NASA, 69
Conclusion and Recommendation, 71

3 MANAGING RISKS TO ASTRONAUT HEALTH 75
General Principles and Issues, 76
Medical Events in Extreme Environments, 78
Health Risk Assessment, 86
Nutrition, 88
Pharmacodynamics and Pharmacokinetics, 90
Environmental and Occupational Health, 92
Health Care Practice Opportunities, 97
Cardiovascular Care, 97
Dental Care, 98

Endocrine Function, 101
Gastrointestinal Issues, 102
Gynecological Health Issues, 103
Hematology, Immunology, and Microbiology, 105
Mental Health Issues, 106
Neurological Issues, 110
Urinary Disorders, 111
Conclusion and Recommendation, 112

4 EMERGENCY AND CONTINUING CARE 117
Anesthesia and Pain Management, 118
Airway Management in Space, 118
Anesthetics, 119
Regional Anesthesia, 120
Surgery and Trauma, 121
Physiological Response to Injury, 121
Surgical Skills and Training, 123
Surgical Equipment, 124
Technical Aspects of Surgery, 125
Prevention of Infection, 126
Management of Common Surgical Emergencies, 126
Management of Abscesses and Soft-Tissue Infections, 127
Rehabilitation for Astronauts on Long-Duration Missions, 127
Catastrophic Illness, Death, and End-of-Life Considerations, 130
Personnel and Other Health Care Resources, 131
Development of a Space Medicine Catalog and Database, 132
Conclusion and Recommendation, 133

5 BEHAVIORAL HEALTH AND PERFORMANCE 137
Astronaut Performance and General Living Conditions, 138
Background, 138
Current Practice and Knowledge Base, 138
Requirements for Additional Knowledge, 142
Use of Pairs of Transport Vehicles for Small Groups Traveling Beyond Earth Orbit, 149
Support and Recovery Systems, 149
Background, 149
Current Practice and Knowledge Base, 150
Requirements for Additional Knowledge, 152

Screening, Selection, and Training, 158
Background, 158
Current Practice and Knowledge Base, 159
Requirements for Additional Knowledge, 161
Strategic Research Considerations, 165
Conclusion and Recommendation, 167

6 EXPLORING THE ETHICS OF SPACE MEDICINE 173
Ethical Issues in Clinical Care for Astronauts, 174
Institutional Pressure to Underreport Clinical Signs, Symptoms, and Medical Data, 174
Current Practice Regarding the Confidentiality of Individual Astronaut Medical Data, 175
Earth-Based Analogs for Balancing Medical Confidentiality with Public Health, 177
Justification for Using the Occupational Health Model to Balance Privacy and Safety, 178
Ethical Issues in the Astronaut-Flight Surgeon Relationship, 180
Opportunities for Collecting Medical Data Aboard the International Space Station, 181
Expand the Collection of Occupational Health Data, 181
Change the Process for Review of Clinical Research Protocols in Space Medicine, 182
Suggested Changes in the Approach to Review of Proposed Clinical Research or Data Collection for Astronauts, 183
Ethical Issues and the Special Circumstances of International Crews, 186
Conclusion and Recommendation, 187

7 PLANNING AN INFRASTRUCTURE FOR ASTRONAUT HEALTH CARE 189
Mission and Goals, 189
Organizational Components, 191
Leadership, 192
Critical Elements of the Organizational Framework, 193
Alternative Organizational Frameworks, 195
Organizational Structure to Ensure Astronaut Health and Safety, 197
Conclusion and Recommendation, 198
Systems Development, 199

Policy and Legislation, 201
Funding, 202
Internal Relationships, 202
External Relationships, 203
Operational and Clinical Components, 208
Standard of Care, 210
Medical Care, 210
Medical Informatics, 212
Personnel and Training, 214
Evaluation and Performance Improvement, 216
Conclusion and Recommendation, 218

REFERENCES 221

APPENDIXES

A Background and Methodology 247
Letter from Daniel S. Goldin, Administrator, NASA 261

B Committee and Staff Biographies 263

INDEX 275

LIST OF BOXES, FIGURES, AND TABLES

EXECUTIVE SUMMARY

Box 1 Clinical Research Opportunities for Astronaut Health, 5
Box 2 Health Care Opportunities in Space Medicine, 8
Box 3 Behavioral Health and Performance Research and Development Opportunities, 14
Box 4 The Key Elements in the Committee's Recommendations, 21

1 ASTRONAUT HEALTH BEYOND EARTH ORBIT

Figure 1-1 Trajectory of a human mission to Mars in 2014, 26
Box 1-1 Charge to the Committee, 31
Box 1-2 What Is Space Medicine in Reference to Developing a Vision and Strategy for Astronaut Health?, 34

2 RISKS TO ASTRONAUT HEALTH DURING SPACE TRAVEL

Box 2-1 Some Major Human Physiological Changes Resulting from Extended Travel in Earth Orbit, 39
Figure 2-1 Countermeasure (CM) evolution, 41
Table 2-1 Average Bone Mineral Density Loss on *Mir*, 44
Box 2-2 Altered Sleep Patterns as Example of Multifactorial Problems Arising During Space Travel, 54
Box 2-3 Elements of the Critical Path Roadmap Project, 63
Table 2-2 Critical Path Roadmap Project: Critical Risks, 64
Figure 2-2 Countermeasure (CM) development and evolution, 68
Box 2-4 What Constitutes Clinical Research?, 69
Box 2-5 Clinical Research Opportunities for Astronaut Health, 72

3 MANAGING RISKS TO ASTRONAUT HEALTH

Box 3-1 Major Health and Medical Issues During Spaceflight, 77
Table 3-1 In-Flight Medical Events for U.S. Astronauts During the Space Shuttle Program, 81
Table 3-2 Medical Events Among Seven NASA Astronauts on *Mir*, 82
Table 3-3 Medical Events and Recurrences Among Astronauts of all Nationalities on *Mir*, 83
Table 3-4 Pharmacopoeia Usage During *Mir* Missions, 84
Table 3-5 Incidence of Health Disorders and Medical-Surgical Procedures During 136 Submarine Patrols, 84
Table 3-6 Reasons for 332 Medical Evacuations from All Submarines, U.S. Atlantic Fleet, 1993 to 1996, 85
Table 3-7 ANARE Health Register Illnesses in Antarctica from 1988 to 1997, 85
Box 3-2 Potential Methods of Risk Assessment and Screening, 86
Box 3-3 Breast Cancer as an Example of Risk Assessment in Space Medicine, 88
Box 3-4 "Normal" Findings on Physical Examination in Microgravity, 90
Box 3-5 Advances in Preventive Dentistry, 99
Box 3-6 Health Care Opportunities in Space Medicine, 113

4 EMERGENCY AND CONTINUING CARE

Box 4-1 Health Care Opportunities in Space Medicine, 134

5 BEHAVIORAL HEATH AND PERFORMANCE

Box 5-1 Evidence of Emotional or Social Problems on Short-Duration Missions, 139
Box 5-2 Recovery and Reintegration, 153
Box 5-3 Systematic Multiple Level Observation of Groups, 156
Box 5-4 Potential Uses of Neuroimaging Methods for Astronaut Selection, Training, and Intervention, 163
Box 5-5 Distributed Interactive Simulation, 165
Box 5-6 Behavioral Health and Performance Research and Development Opportunities, 168

6 EXPLORING THE ETHICS OF SPACE MEDICINE

Box 6-1 Examples of Earth-Based Limitations on Doctor-Patient Confidentiality to Decrease Risks to Others, 176
Box 6-2 The Common Rule and Informed Consent, 182
Box 6-3 Potential Categories of Clinical Research or Data Collection Protocols for Astronauts, 184
Box 6-4 Examples of Category 2 Clinical Research or Data Collection Protocols for Astronauts, 184

7 PLANNING AN INFRASTRUCTURE FOR ASTRONAUT HEALTH CARE

Box 7-1 Infrastructure Elements for Developing a Comprehensive Health Care System for Astronauts, 199

SAFE PASSAGE

Astronaut Care for Exploration Missions

Executive Summary

ABSTRACT

Space travel is inherently risky. Space beyond Earth orbit is an extreme and isolated unique environment. Currently, not enough is known of the risks of prolonged travel in deep space to enable humans to venture there for prolonged periods safely. To support safe human exploration of space, the National Aeronautics and Space Administration (NASA) should pursue a two-component strategy: (1) it should pursue a comprehensive health care system for astronauts to capture all relevant epidemiological data, and (2) it should pursue a long-term, focused health care research strategy to capture all necessary data on health risks and their amelioration. An occupational health model should apply to the first pursuit, and a modification of the interpretation of the Common Rule (45 C.F.R., Part 46, Subpart A) for human research participants should apply to the second one. One special focus of research should be the complex behavioral interactions of humans in extreme, isolated microenvironments such as inside spacecraft. To accomplish this strategy, there should be an organizational component within NASA that has authority over and accountability for all aspects of astronaut health.

Space travel is inherently risky, and space travel on long-duration missions (those of a year or longer) beyond Earth orbit (beyond the orbital band of launched satellites and the International Space Station [ISS]) entails special risks to humans. Deep space is a unique environment. It is unique for several reasons: (1) it likely has unknown risks, (2) there are no validated effective responses to most of the known risks that humans will encounter there, and (3) it isolates humans, in that humans in deep space will not have the capability for either real-time communication with Earth or a timely return. The acquisition of a fundamental understanding of these risks and the development of solutions to the problems that they present are the subjects of this report.

TASK OF THE COMMITTEE

The general charge to the Committee on Creating a Vision for Space Medicine During Travel Beyond Earth Orbit was to develop a vision for space medicine for long-duration space travel. With the important exception of the ISS, such travel is many years in the future. During the interim, innumerable changes will occur, many of which are unpredictable. As new knowledge is developed, humans will learn much that is directly applicable to the task of enabling safe space travel. Institutional arrangements will shift, and as priorities change, new management principles will be applied; often, these will be affected by political realities. In this report, the committee focuses on the development of principles that should guide future approaches to the issues.

In planning for long-duration space travel beyond Earth orbit, the National Aeronautics and Space Administration (NASA) is undergoing a transition from the relatively known (e.g., the space shuttle has flown more than 100 missions) to the unknown. In addition, a second transition is occurring: from an emphasis on the machinery of spaceflight to an increased emphasis on the biology of spaceflight. For both NASA and the engineering community this is a conceptual shift that has important practical implications as biology adds to chemistry, mathematics, and physics as guiding sciences in engineering in general and in NASA's mission. The challenges afforded by these twin transitions offer NASA a strategic opportunity to reexamine its processes and structure and to build on its successes.

In addition to the general charge to the committee, NASA gave the committee several specific tasks. Chapter 1 introduces the health problems that may confront humans in deep space. Chapter 2 addresses what is known about the risks to health during space travel and where clinical research opportunities exist. Chapters 3 and 4 review what is known about health care during space travel and where opportunities may exist for the development of effective approaches to health care during travel in deep space. Chapters 5 and 6 highlight two specific areas that the committee believes are critical: (1) behavioral, cultural, and social issues (Chapter 5) and (2) an approach to the collection of the clinical data necessary to ensure the safety of space travel beyond Earth orbit (Chapter 6). Chapter 7 suggests ways in which an effective health care system for astronauts might be organized.

Two themes run throughout the report: (1) that not enough is yet known about the risks to human health during long-duration missions beyond Earth orbit or about what can effectively mitigate those risks to enable humans to travel and work safely in the environment of deep space and (2) that everything reasonable should be done to gain the necessary information before humans are sent on missions of space exploration.

RISK

Throughout its history, NASA has dealt successfully with transition: the transition from atmospheric flight, to supersonic flight, to suborbital flight, to orbital flight (which culminated with the space shuttle), and to orbital missions, both with *Mir* and, more recently, with the ISS. Long-duration missions beyond Earth orbit represent another transition and another opportunity. Such missions are not merely quantitatively different; they are also qualitatively different.

The three most important health issues that have been identified for long-duration missions are radiation, loss of bone mineral density, and behavioral adaptation. First, although exposure to radiation is of concern during missions in low Earth orbit, its potential effects become more acutely worrisome during extravehicular activity and are chronically worrisome for those living on the ISS. Longer-duration missions increase the risk at least arithmetically because of the length of the mission and the changing character of radiation in the environment. This is a formidable challenge for engineering, basic biomedical, and clinical research, as discussed in the section Environmental and Occupational Health in Chapter 3. Second, loss of bone mineral density, which apparently occurs at an average rate of 1 percent per month in microgravity, is relatively manageable on the short-duration missions of the space shuttle, but it becomes problematic on the ISS, as described in Chapter 2. If this loss is not mitigated, interplanetary missions will be impossible. Finally, human interactions aboard a spacecraft, isolated in time and space from Earth, may well be one of the more serious challenges to exploratory missions by humans (Chapter 5).

Risk is of high priority to NASA, and determination of what risks to humans exist and what countermeasures should be taken are addressed through NASA's Critical Path Roadmap project ("countermeasure" is NASA's designation for preventive and therapeutic interventions before or during space missions). Nevertheless, risk should be addressed at other levels. At the level of the individual astronaut, for example, how may an astronaut come to a personal decision in a truly informed way to accept the risk of a maiden voyage to Mars? To make an informed personal decision, astronauts should be involved in the process that identifies risks and their amelioration, not only from the standpoint of immediate countermeasures over which they might have control while in flight but also from the standpoint of those risks for which they have no personal or immediate control. At the level of society, on the other hand, risks should be addressed explicitly. The successes of the space program may have fostered the impression that space travel has few associated risks. Making potential problems and overall risks clear and openly disclosing them will allow NASA to gain continuing public understanding, trust, and support for exploration-class space missions. NASA can tailor the amount of detail disclosed in relation to the anticipated severity and prevalence of the risks to astronaut health and safety and the level of support that NASA is seeking. At the extreme, the public must be prepared for the possibility that all countermeasures may tragically fail, that a crew may not return from a prolonged mission, or that individuals may not be able to function physically or mentally upon their return.

There is a profound professional and ethical responsibility to evaluate honestly the risk to human life that accompanies long-duration space travel. This risk should be evaluated through clinical research (Box 1) in the context of the benefit to humans, but it should be stated at the level of the individual in terms that can be plainly understood.

Conclusion 1

Space travel is inherently hazardous. The risks to human health of long-duration missions beyond Earth orbit, if not solved, represent the greatest challenge to human exploration of deep space. The development of solutions is complicated by lack of a full understanding of the nature of the risks and their fundamental causes.

- *The unique environment of deep space presents challenges that are both qualitatively and quantitatively different from those encountered in Earth orbit. Risks are compounded by the impossibility of a timely*

BOX 1
Clinical Research Opportunities for Astronaut Health

Musculoskeletal System

1. Establishing the course of changes in bone mineral density and markers of bone mineral density turnover in serum and urine before, during, and after space travel.
2. Developing a capacity for real-time measurement of bone mineral density and enhanced three-dimensional technology to assess the risk of fracture during space travel.
3. Identifying human phenotypes and genotypes resistant to space travel-induced bone mineral density loss.
4. Tailoring therapeutic interventions (i.e., countermeasures such as diet, exercise, and medications) as a high priority and then validating the promising countermeasures in studies with astronauts during exposure to microgravity.

Cardiovascular

5. Considering artificial gravity and pharmacological interventions as solutions to orthostatic hypotension.

Gastrointestinal

6. Investigating the relationship between space motion sickness and absent bowel sounds including pharmacological and adaptive countermeasures.

Nervous System

7. Building a coordinated clinical research program that addresses the issues of neurological safety and care for astronauts during long-duration space travel.
8. Performing pharmacological trials with dose-response and pharmacokinetic measures to assess the efficacies and toxicities of medications commonly used to treat sleep disturbances during space travel.
9. Considering clinical trials on the use of growth hormone or other countermeasures and developing devices to control spacecraft ambient light and the core temperature at appropriate levels during space travel to reduce sleep disturbances.

Reproductive Health

10. Determining whether radiation exposure during space travel causes genetic damage or altered fertility in men and women and, for women, premature ovarian failure.
11. Determining female and male reproductive hormone levels during space travel.
12. Determining the effect of microgravity on menstrual efflux and retrograde menstruation.

Physiological Monitoring

13. Collecting clinical data for both men and women when anatomically possible and physiologically sensible for all individuals in the space program and, on a regular basis, subjecting the data to analysis for sex-related differences.
14. Giving priority to high-resolution, high-precision, yet minimally invasive or noninvasive methods for the monitoring of physiological parameters and for imaging of the human body during space travel.

return to Earth and of easy resupply and by the greatly altered communications with Earth.

- *The successes of short-duration space missions may have led to misunderstanding of the true risks of space travel by the public. Public understanding is necessary both for support of long-duration missions and in the event of a catastrophe.*

Recommendation 1

NASA should give increased priority to understanding, mitigating, and communicating to the public the health risks of long-duration missions beyond Earth orbit.

- **The process of understanding and mitigating health risks should be open and shared with both the national and the international general biomedical and health care research communities.**
- **The benefits and risks—including the possibility of a catastrophic illness or death—of exploratory missions should be communicated clearly, both to astronauts and to the public.**

HEALTH CARE

To understand, prevent, and mitigate risks, knowledge of the risks is necessary. Because of the relatively few opportunities to acquire and analyze data from studies conducted in microgravity environments, every possible opportunity to do so must be exploited. Opportunities for the collection of two types of data exist: clinical data on the astronauts and the results of astronaut health care research (Box 2).

Clinical data, including personal health data, have been collected over the 40 years that humans have flown in space, but data collection has not been done in a systematic way, nor have the data been fully analyzed. A comprehensive health care system for astronauts—both active and retired—should ensure that all data relevant to space travel are collected. Combined with a strategic health care research plan that would enable the analysis of those data, such a system would foster data-driven decisions about health risks, prevention, and mitigation.

"Comprehensive" means that all health care for astronauts is coordinated through an astronaut health care system and covers all periods while the astronaut is active, including the selection, premission, intramission, postmission, and intermission phases. "Comprehensive" also means that there is retrospective as well as prospective collection and analysis of clinical data. The astronaut health care system should include not only a health care component but also health care research and training components. The standard of clinical care for a health care system for astronauts should be equivalent to the best clinical care available on Earth for those problems that occur before and after a mission. The goals of the health care system should be to maximize the astronaut's ability to function as a productive member of the crew while in deep space and to maintain or to restore normal function in the premission and the postmission phases.

Conclusion 2

Crew health has not received the attention that it must receive to ensure the safety of astronauts on long-duration missions beyond Earth orbit, nor has NASA sufficiently integrated astronaut health care into mission operations.

- *Currently, there is no comprehensive and inclusive strategy to provide optimum health care for astronauts in support of long-duration missions beyond Earth orbit, nor is there sufficient coordination of health care needs with the engineering aspects of such missions.*
- *An effective health care system is founded on data that are accumulated, analyzed, and used to continuously improve health care for astronauts on future space missions. Inherent in an appropriate health care system is a mechanism that can be used to gather and analyze data relevant to key variables. NASA could have collected and analyzed many more medical data had a comprehensive health care system focused on astronauts been in place and been given the priority and resources that it needed.*
- *Although the equipment and expertise that will be needed to provide health care during future long-duration missions beyond Earth orbit cannot be reliably predicted, a health care system that is data driven and linked to a research strategy will position NASA to better monitor pertinent developments and meet future challenges.*

Recommendation 2

NASA should develop a comprehensive health care system for astronauts for the purpose of collecting and analyzing data while providing the full continuum of health care to ensure astronaut health. A NASA-sponsored health care system for astronauts should

BOX 2
Health Care Opportunities in Space Medicine

1. Expanding, validating, and standardizing a modified physical examination, the microgravity examination technique, and including a technique for pelvic examination for use in microgravity.
2. Developing an easily identifiable database for medications on the spacecraft, including dosage, indications, adverse effects, and anticipated changes in the pharmacokinetic profile in microgravity.
3. Developing an easily accessible hazardous materials manual for space travel to aid in the surveillance, detection, decontamination, and treatment of chemical exposures.
4. Monitoring and quantifying particulates on a continuing basis.
5. Examining the capability of microbial identification, control, and treatment during space travel.
6. Developing methods for noise cancellation or reduction.
7. Standardizing ergonomic practices on the basis of the human body's response to the microgravity environment.
8. Developing methods to measure human solar and cosmic radiation exposures and the means to prevent or mitigate their effects.
9. Providing a thorough cardiovascular evaluation similar to the premission evaluation at the cessation of space travel to provide useful data as part of the continuum of astronaut care and to aid in establishing an evidence base for cardiovascular disorders during space travel.
10. Developing a program for instruction in basic dental prophylaxis; the treatment of common dental emergencies such as gingivitis, tooth fracture, dental trauma, caries, and dental abscesses; and tooth extractions.

- **care for current astronauts, astronauts who are in training, and former astronauts, as well as, where appropriate, their families;**
- **cover all premission, intramission, and postmission aspects of space travel;**
- **incorporate innovative technologies and practices—including clinical practice guidelines—into prevention, diagnosis, treatment, and rehabilitation, including provision for medical care during catastrophic events and their sequelae;**
- **be uniform across the international space community and cooperatively developed with the international space community; and**
- **receive external oversight and guidance from prominent experts in clinical medicine.**

RESEARCH

The goal of NASA-sponsored health care research is, first, to learn how

11. Studying the bioavailability and pharmacological function of exogenous hormone therapy during space travel and, as new medical therapies for gynecological surgical conditions evolve, testing these therapies for use during space travel.

12. Performing clinical studies on anemia, immunity, wound infection, and wound healing as part of every space mission.

13. Developing methods for the identification and management of mood disorders and suicidal or homicidal ideation and developing protocols for the management of violent behavior, including crisis intervention, pharmacological restraint, and physical restraint.

14. Establishing a coordinated clinical research program that addresses the issues of neurological safety and care for the astronauts during long-duration missions beyond Earth orbit.

15. Developing a resource-based medical triage system that contains guidelines for the management of individual and multiple casualties during space travel.

16. Developing an anesthetic approach associated with rapid and comfortable recovery using anesthetic drugs with short durations of action or for which there are antagonists.

17. Creating guidelines for withdrawal of care in space and for dealing with the death of a crewmember from physiological and behavioral points of view.

18. Developing a mechanism for skill maintenance and retraining in psychomotor skills during long-duration space missions.

19. Recording routine surveillance of health status measures, incidents of illnesses and injuries, and their treatments in a database with standardized rates of occurrence so that data between studies and missions can be compared.

20. Developing and maintaining a centralized catalogue of all written materials related to space and analog-environment biomedical research and experience according to current medical informatics standards.

to send and keep humans safely in space and have them return to Earth in good health. In Chapter 7 of this report, the committee recommends a comprehensive health care system that will enable the collection of all relevant clinical information. Such clinical information—the results of natural experiments that reveal common physiological responses to the microgravity environment—may be thought of as epidemiological data. The results of targeted and planned experimentation provide a second element of the health care research plan, broadly construed. Chapter 2 reviews what is known from both epidemiological and targeted research and suggests opportunities for further research.

The principle underlying a health care research strategy is that it have a steadfast, prospective, and methodologically sound approach to the collection of data. Although some work on assessing the efficacies of countermeasures has been done, none has been shown to be effective in reducing the most significant effects of microgravity (bone mineral density loss, muscle loss, and neurovestibular maladaptation). Although artificial and analog environments on Earth have been useful in predicting certain effects of microgravity and isolation on humans, there is no substitute for the microgravity environment for clinical research. The ISS represents the single most important test bed for that research.

Both the intramural and extramural research programs of NASA, especially the research conducted through the National Space Biomedical Research Institute, are mechanisms for addressing two principal issues: the relative paucity of relevant clinical data and the heretofore relatively closed clinical research environment that has characterized NASA. NASA should have a process that opens its research awards and requests for proposals to the widest audience, and NASA should foster substantial investigator-initiated research. A far more basic understanding of fundamental processes is needed, however; and investigator-initiated research in broad areas is an appropriate strategy for the development of such an understanding.

Whatever the organizational approach, the outcome should be an infrastructure that will (1) test hypotheses about the risks to humans during long-duration travel beyond low Earth orbit, (2) test the efficacies of treatments or countermeasures in reducing or eliminating the defined risks, and (3) evaluate the long-term sequelae of space travel on humans.

Conclusion 3

NASA has devoted insufficient resources to developing and assessing the fundamental clinical information necessary for the safety of humans on long-duration missions beyond Earth orbit.

- *Although humans have flown in space for nearly four decades, a paucity of useful clinical data have been collected and analyzed. The reasons for this include inadequate funding; competing mission priorities; and insufficient attention to research, analysis including insufficient investigator access to data and biological samples, and the scientific method.*
- *Although NASA's current approach to addressing health issues through the use of engineering design and countermeasures has been successful for short-duration missions, deep space is a unique environment that requires a different approach.*
- *A major problem of space medicine research is the small number of astronaut research participants, which requires special design and analysis of the data from clinical trials with small numbers of participants. This necessitates a strategy focused on maximization of opportunities for learning.*

Recommendation 3

NASA should develop a strategic health care research plan designed to increase the knowledge base about the risks to humans and their physiological and psychological adaptations to long-duration space travel; the pathophysiology of changes associated with environmental forces and disease processes in space; prediction, development, and validation of preventive, diagnostic, therapeutic, and rehabilitative measures for pathophysiological changes including those that are associated with aging; and the care of astronauts during space missions.

The strategic research plan should be systematic, prospective, comprehensive, periodically reviewed and revised, and transparent to the astronauts, the research community, and the public. It should focus on

- **providing an understanding of basic pathophysiological mechanisms by a systems approach;**
- **using the International Space Station as the primary test bed for fundamental and human-based biological and behavioral research;**
- **using more extensively analog environments that already exist and that have yet to be developed;**
- **using the research strengths of the federal government, universities, and industry, including pharmaceutical, bioengineering, medical device, and biotechnology firms; and**
- **developing the health care system for astronauts as a research database.**

BEHAVIORAL HEALTH

The prototype for a long-duration mission beyond Earth orbit is an interplanetary mission, with Mars as the likely destination. Such a voyage of discovery and return will take nearly 3 years. The crew is likely to be multicultural, international, and of both sexes. They will spend all their time together in a very confined space. The habitability of the spacecraft will be compromised by the need to carry all necessary equipment and nourishment (even if it is replenishable), at least for the voyage to Mars (if supplies have already been stockpiled there), if not for the entire 3 years. Real-time communication with Earth will be impossible, as at the farthest distance from Earth, radio messages and messages transmitted by even more advanced means will take 20 minutes to reach their destination. Finally, the success of the mission and the lives of the astronauts may depend on every member of the crew functioning appropriately, both physically and emotionally.

Maintaining a healthy behavioral condition, that is, behavioral health, in such an extreme, isolated microenvironment is, in the committee's view, critical to the success of the mission and to the return to Earth of normally functioning human beings. Understanding the behavior of individuals in such environments and understanding the interactions of members of a team in prolonged isolation are necessary prerequisites to preventing disruptive behaviors and to dealing with them adequately should they occur. Crew selection and training are also key in the process, but much remains unknown (Box 3).

NASA has taken steps in two areas of importance: it has learned from the experiences of others on missions in analog environments such as operations on research stations in Antarctica, prolonged submarine missions, and deep diving operations; and it has learned about the nature of teams and the interaction of their members. The distinction between missions in low Earth orbit and long-duration missions beyond Earth orbit is that in the latter there is no means of timely return, and all problems, including disruptive behaviors and negative crew interactions, must be dealt with within the spacecraft by the crew.

Historically, crew selection has been an opaque process, in terms both of the individuals selected and of the composition of the crew. Both crew selection and crew composition will assume far greater importance, as the compatibility of individuals has the potential of preventing, to a degree, significant disruptive behaviors. Crew training in positive interactions and conflict resolution, as well as in dealing with adverse behavioral events, will be necessary.

Conclusion 4

Behavioral health and performance effectiveness present major challenges to the success of missions that involve quantum increases in the time and the distance traveled beyond Earth orbit.

- *The available evidence-based spaceflight data are insufficient to make an objective evaluation or projection regarding the behavioral health issues that are likely to arise.*
- *The analysis of the complex individual and group habitability interactions that critically influence behavioral health and performance*

effectiveness in the course of long-duration missions remains to be planned and undertaken.

- *There is a need for more information about support delivery systems at the interface between ground-based and space-dwelling groups.*
- *In the absence of a valid and reliable analysis of the existing database, it is not possible to determine whether the current procedures will be adequate for the screening and selection of candidates for long-duration missions.*
- *Although the data from natural analog environments, including simulation studies, may be helpful, there remains a need to accumulate knowledge based on observations from systematic research in both natural and simulated extreme terrestrial environments and venues like the International Space Station.*

Recommendation 4

NASA should give priority to increasing the knowledge base of the effects of living conditions and behavioral interactions on the health and performance of individuals and groups involved in long-duration missions beyond Earth orbit. Attention should focus on

- **understanding group interactions in extreme, confined, and isolated microenvironments;**
- **understanding the roles of sex, ethnicity, culture, and other human factors on performance;**
- **understanding potentially disruptive behaviors;**
- **developing means of behavior monitoring and interventions;**
- **developing evidence-based criteria for reliable means of crew selection and training and for the management of harmonious and productive crew interactions; and**
- **training of both space-dwelling and ground-based support groups specifically selected for involvement in operations beyond Earth orbit.**

DATA COLLECTION AND ACCESS

Because of the high degree of risk of long-duration space missions and the relatively few data available, the need for the collection and analysis of all relevant data is a message that appears throughout this report. Currently, NASA distinguishes between astronaut health-related data, which are medical data and which are considered private, and supplemental data (mission-based data and responses to space travel) and integrated test regimen data

BOX 3
Behavioral Health and Performance Research and Development Opportunities

Astronaut Performance, General Living Conditions, and Group Interactions

1. Enhancing the evidence base on the organization of general living conditions and performance requirements for small groups of humans in isolated and confined microsocieties over extended time intervals and developing an evidence-based approach to the management of harmonious and productive, small, multinational groups whose members will have to function effectively in isolated, confined, and hazardous environments.

2. Coordinating the development of design engineering and habitability requirements on the one hand and evidence-based behavioral health imperatives on the other.

3. Identifying and analyzing those features of small social systems that foster the effectiveness of groups functioning semiautonomously over extended periods of time.

4. Analyzing potentially disruptive group influences that adversely affect harmonious and productive performance interactions under the isolated, confined, and hazardous conditions that characterize long-duration space missions beyond Earth orbit.

Support and Recovery Systems

5. Developing a technology that will provide an adequate means for assessment of the behavioral health effects of long-duration space missions and that will establish and maintain safe and productive human performance in isolated, confined, and hazardous environments and developing an evidence-based approach to the establishment and maintenance of a system for the delivery of behavioral health support and to the analysis of those internal and external factors that influence the effectiveness of the system.

6. Evaluating and enhancing communication with family, friends, and other ground personnel and onboard recreational activities as means of providing behavioral health support for long-duration missions.

7. Evaluating the validity and reliability of performance-monitoring procedures including the Crew Status and Support Tracker, the Windows Space Flight Cognitive Assessment Tool, and the Space Behavioral Assessment Tool and the extent to which methodologies for intramission performance monitoring are enhanced by online downlink capabilities in studies conducted during long-duration missions.

8. Developing and refining procedures for effective intervention under conditions of potentially disruptive personal interactions both among astronauts and between astronauts and Earth-bound support components and for evaluation of the nature and extent of changes in group interaction patterns.

9. Evaluating the effects of ground-based support system design factors including backup components and personnel changes on group integration and stability as they affect personal coaching and technical support functions.

10. Developing and assessing countermeasure interventions that meet the challenges presented by emergencies and technical assistance requirements under conditions with complexities related to cultural and language differences as well as under conditions that involve crews of mixed sexes.

Screening, Selection, and Training

11. Systematically analyzing and evaluating the extensive existing database on the methods and procedures for screening and selection of astronauts used over the past several decades.

12. Evaluating personality measures in the development of valid and reliable procedures for the screening and selection of astronauts and determining the extent to which intelligence and aptitude measures may predict performance more accurately than the more commonly applied personality measures.

13. Developing and evaluating screening and selection procedures that validly and reliably discriminate effective group interaction skills and competences.

14. Developing and refining training technologies including automated training for the preparation of multinational space-dwelling microsocieties as well as their Earth-bound support groups including those related to the interactions of individuals and small groups in the context of distributed ground-based and space-dwelling performance sites.

Data Collection, Analysis, and Monitoring

15. Incorporating and enhancing relevant behavioral health factors as an effective contribution to a more comprehensive plan for the collection and management of astronaut health care data.

16. Developing and testing valid and reliable individualized monitoring and assessment procedures to enhance intrapersonal self-management.

17. Refining communication-monitoring techniques and countermeasure interventions for interactions within and between ground-based and space-dwelling groups.

18. Developing and refining technological approaches to the assessment of individual and group behavioral integrity as well as the efficacies of countermeasure evaluations during long-duration space missions.

19. Establishing a systematic approach to the collection and analysis of postmission (recovery), debriefing, and longitudinal follow-up astronaut health data, including data on behavioral health and performance components.

20. Planning and undertaking a systematic collation and relational analysis of the existing archives of astronaut evaluation data to develop an evidence-based approach to valid and reliable means of screening and selection of candidates for long-duration missions beyond Earth orbit.

(research data from studies with astronauts as participants), which are available for analysis, although they are nonattributable. The reality is that the astronaut is, in most cases, the only individual from whom clinical information relevant to space travel can be collected. Therefore, reliance on the voluntary participation of astronauts in clinical research to the same extent as reliance on volunteer participants on the ground may not be appropriate. This is especially true when the information gained is potentially critical to the lives and well-beings of both the individual astronaut and the astronaut corps.

The crux of the issue is this: is the astronaut the same as any other volunteer participant of human experimentation, or is he or she de facto an experimental participant? More importantly, is there a middle ground, one that ensures the collection of all relevant data while protecting the individual from the inappropriate release of private information?

Conclusion 5

The ultimate reason for the collection and analysis of astronaut health-related data is to ensure the health and safety of the astronauts.

- *Emphasis on the confidentiality of astronaut clinical data has resulted in lost opportunities to understand human physiological adaptations to space, and concern for the protection of privacy and over the implications regarding disclosure and use of clinical data may have led to the underreporting of relevant information.*
- *Reevaluation of the application of the Privacy Act and statutory privacy provisions may be necessary to enable appropriate access to necessary data while protecting the privacy of the individual astronaut.*
- *The unique environment of deep space, combined with the social and institutional contexts of health care research with astronauts, requires that astronauts be considered a unique population of research participants.*
- *A limited international consensus exists on the appropriate principles and procedures for the collection and analysis of astronaut medical data. The potential for conflict among the national space agencies and International Space Station partners is high.*

Recommendation 5

NASA should develop and use an occupational health model for the collection and analysis of astronaut health data, giving priority to the creation and maintenance of a safe work environment.

- **NASA should develop new rules for human research participant protection that address mission selection, the limited opportunities for research on human health in microgravity, and the unique risks and benefits of travel beyond Earth orbit.**
- **A new interpretation or middle ground in the application of the Common Rule (45 C.F.R., Part 46, Subpart A) to research with astronauts is needed to ensure the development of a safe working environment for long-duration space travel.**
- **NASA should continue to pursue consensus among national space agencies and International Space Station partners on principles and procedures for the collection and analysis of astronaut medical data.**

ENGINEERING AND BIOLOGY

The preceding section and Chapter 6 of this report outline a conceptual shift in terms of the collection and analysis of astronaut health data using an occupational health model. A second shift is the movement of engineering and biology toward each other.

The approaches to problem solving that NASA historically applied to engineering design and machinery may not be appropriate when they are applied to human anatomy and physiology. In engineering terms, NASA identifies risk and then develops countermeasures. The development of a specific countermeasure may be addressed by a task order to a contractor. In biological terms, however, the degree of risk varies because of biodiversity. What this implies practically is that the task order approach, likely appropriate as a solution to a technological problem, may be insufficient as a solution to a biological problem; a broader approach, one that develops a deeper understanding of the nature and causes of risk and the diversity of the responses, may be necessary. This broader approach is reflected in the committee's recommendation for a strategic health care research plan for astronauts.

NASA originated because of an engineering need, and its most remarkable successes have been technological. Indeed, the committee heard in its discussions with astronauts that even the most biologically oriented of astronauts, the physician-astronauts, frequently identify themselves first as astronauts and then as physicians. A conceptual shift from engineering to biology, however, must take into account the fact that people are not machines. Such a shift must also take into account the fact that the habitat and environment that may be appropriate for machines on long-duration missions or even for humans on short-duration missions may be inappropriate or even detrimental to the productivity and well-being of humans on long-duration missions into deep space.

Conclusion 6

Exploratory missions with humans involve a high degree of human-machine interaction. The human factor will become more important as the durations of missions into deep space with humans increase and as the spacecraft crew functions more autonomously, adapts to unexpected situations, and makes real-time decisions.

- *NASA, because of its mission and history, has tended to be an insular organization dominated by traditional engineering. Because of the engineering problems associated with early space endeavors, the historical approach to solving problems has been that of engineering. Long-duration space travel will require a different approach, one requiring wider participation of those with expertise in divergent, emerging, and evolving fields. NASA has only recently begun to recognize this insufficiency and to reach out to communities, both domestic and international, to gain expertise on how to remedy it.*
- *Engineering and biology are increasingly integrated at NASA, and this integration will be of benefit to the flexibility and control of long-duration missions into deep space. NASA's structure does not, however, easily support the rapidly advancing integration of engineering and biology that is occurring throughout the engineering world outside NASA. NASA does not have a single entity that has authority over all aspects of astronaut health, health care, habitability, and safety that could facilitate the integration of astronaut health and health care with engineering.*
- *The human being must be integrated into the space mission in the same way in which all other aspects of the mission are integrated. A comprehensive organizational and functional strategy is needed to coordinate engineering and human needs.*

Recommendation 6

NASA should accelerate integration of its engineering and health sciences cultures.

- **Human habitability should become a priority in the engineering aspects of the space mission, including the design of spacecraft.**
- **Investigators in engineering and biology should continue to ex-**

plore together and embrace emerging technologies that incorporate appropriate advances in biotechnology, nanotechnology, spaceworthy medical devices, "smart" systems, medical informatics, information technology, and other areas to provide a safe and healthy environment for the space crew.

- **More partnerships in this area of integration of engineering and health sciences should be made with industry, academic institutions, and agencies of the federal government.**

ORGANIZATION

Throughout this report the committee expresses concern over the lack of existing data and the lack of analyses of data on which decisions about the health and safety of astronauts on long-duration missions beyond Earth orbit can be based. For that reason, the committee recommends two basic mechanisms for retrieval of the necessary information: (1) a health care system for astronauts that will allow the collection and analysis of epidemiological data and (2) a strategic health care research plan that will promote the development of a fuller understanding of the risks of space travel to human health. The committee further recommends that NASA consider a different model—that of occupational health—for use in its approach to data collection. Implementation of these conceptual and practical recommendations will require a different structure within NASA. An approach to planning the organizational and administrative structures needed to ensure astronaut health is presented in Chapter 7.

Conclusion 7

The challenges to humans who venture beyond Earth orbit are complex because of both the unique environment that deep space represents and the unsolved engineering and human health problems related to long-duration missions in deep space. The committee believes that the organizational structure of NASA may not be appropriate to successfully meet the challenge of ensuring the health and safety of humans on long-duration missions beyond Earth orbit.

- *Astronaut health and performance will be central to the success of long-duration space missions, but the responsibility for astronaut health and performance is buried deep within NASA.*
- *Within NASA the focus on health care research and astronaut health care is not sufficient, nor does NASA sufficiently coordinate and*

integrate the research activities needed to support successful long-duration missions beyond Earth orbit.

Recommendation 7

NASA should establish an organizational component headed by an official who has authority over and accountability for all aspects of astronaut health, including appropriate policy-making, operational and budgetary authority. The organizational component should be located at an appropriate place and level in the NASA organizational structure so that it can exercise the necessary authority and responsibility. The official who heads the organizational component should be assisted by officials who are separately responsible for astronaut clinical care and health care research. The proposed organizational component should

- **have authority over basic, translational, and clinical biomedical and behavioral health research;**
- **foster coordination between NASA and the external research community; and**
- **be overseen by an external advisory group, modeled on advisory groups of the National Institutes of Health and other federal external advisory groups, to provide program review, strategic planning, and leverage to assist NASA in meeting its goals for astronaut health.**

BOX 4
The Key Elements in the Committee's Recommendations

Managing and Communicating Risks to Astronaut Health

Recommendation 1. NASA should give increased priority to understanding, mitigating, and communicating to the public the health risks of long-duration missions beyond Earth orbit.

Comprehensive Astronaut Health Care System

Recommendation 2. NASA should develop a comprehensive health care system for astronauts for the purpose of collecting and analyzing data while providing the full continuum of health care to ensure astronaut health.

Strategic Health Care Research Plan

Recommendation 3. NASA should develop a strategic health care research plan designed to increase the knowledge base about the risks to astronaut health.

Understanding Behavioral, Social and Cultural Issues and Challenges

Recommendation 4. NASA should give priority to increasing the knowledge base of the effects of living conditions and behavioral interactions on the health and performance of astronauts on long-duration space missions.

Astronaut Health and Safety Data Collection and Access

Recommendation 5. NASA should develop and use an occupational health model for the collection and analysis of astronaut health data, giving priority to the creation and maintenance of a safe work environment.

Integration of Engineering and Health Sciences

Recommendation 6. NASA should accelerate integration of its engineering and health sciences cultures.

Authority and Accountability for Astronaut Health

Recommendation 7. NASA should establish an organizational component headed by an official who has authority over and accountability for all aspects of astronaut health, including appropriate policy-making, operational, and budgetary authority.

Landing of the space shuttle *Discovery* at the Kennedy Space Center on June 12, 1998, marking the end of STS-91, the final space shuttle-*Mir* docking mission, and 812 days of continuous U.S. presence in Earth orbit. NASA image.

1
Astronaut Health Beyond Earth Orbit

. . . if any man could arrive at the exterior limit [deepest space], . . . he would see a world [universe] beyond [the Earth]; and, if the nature of man could sustain the sight, he would acknowledge that this other world [universe] was the place of the true heaven and the true light and the true earth.

Socrates, c400 B.C. (from Plato's Dialog Phaedo [109e])

BACKGROUND

For more than three decades the U.S. space community has been planning to send humans on exploration-class missions to Mars (Burrows, 1998). In the post-Apollo space mission era, this would be the next "giant leap for mankind" and the first step toward human exploration of the solar system. Although Mars is a cold and inhospitable place with an extreme environment and an atmosphere with high levels of carbon dioxide that cannot support human life, it is also the nearest and most Earth-like planet in the solar system. Mars has seasons, polar ice caps, mountains and canyons, volcanoes, and evidence of ancient rivers and lakes. It is the most accessible body among the planets and moons in the solar system where a sustained human presence is believed to be possible (Hoffman and Kaplan, 1997).

At the time of the public release of this report, there are no concrete plans to land humans on Mars; that event is more than a decade in the future. However, much of the engineering research and logistical planning needed to make such a mission a reality has been under way for years and continues today in many fields and on many fronts. The most recent demonstration of the human insatiable desire to explore and conquer the "outer limits" of habitability was the ascent on October 31, 2000, of Expedition 1 from a launchpad in Kazakhstan, which carried the first international astronaut-cosmonaut contingent to commence habitation of the International Space Station and a permanent human presence in Earth orbit.

Since 1991 the U.S. National Aeronautics and Space Administration (NASA) has been studying how to make human exploration of Mars a "feasible undertaking for the space-faring nations of Earth" (Hoffman and Kaplan, 1997, Section 1, p. 3). The agency published the cumulative results of its thinking in *Human Exploration of Mars: The Reference Mission of the NASA Mars Exploration Study Team* (Hoffman and Kaplan, 1997). The report characterized the human exploration of Mars as a goal that "currently lies at the ragged edge of achievability" (Hoffman and Kaplan, 1997, Section 1, p. 5). Human exploration of this scope requires optimum functioning of both spacecraft and astronauts—of both the engineering and the human components. Failure of either could result in mission failure. Success thus requires close integration of both throughout the design, planning, and implementation process.

Of all the challenges that such missions beyond Earth orbit imply, the most daunting will be to provide for the health and safety of astronauts who venture beyond Earth's orbit for the first time. Their well-being will depend in part on future advances in medicine and engineering (SSB and NRC, 1996, 1998a,d). It will also depend upon biologically inspired technologies under development, such as nanosensors that can monitor health status, and quantum advances in informatics and robotics. On the surface of Mars, astronauts will rely for safety on advance teams of so-called smart robots that can "do the dangerous work and keep our astronauts out of harm's way" (Goldin, 1999).

Keeping astronauts out of "harm's way" also means designing a health care system that is contemporary, practical, and portable. The system must be grounded in clinical evidence, yet it must take into account both risks that are known and predictable and those that have yet to be determined. Finally, the health care system must work in a setting that is far more remote and far more extreme in physical and other environmental conditions than anything modern health planners, practitioners, and patients have yet encountered, where neither timely resupply nor timely return to the point of departure will be possible.

How to proceed toward designing a contemporary, practical, and portable health care diagnosis and delivery system to successfully meet this future necessity is the subject of this report. The challenge of designing such a system is underscored by the observations made thus far: that space travel is severely debilitating to humans in many ways (SSB and NRC, 1998a, 2000). Recovery from these extreme physiological changes is uncertain and may be incomplete at best. Moreover, NASA believes that simply landing humans on Mars and returning them safely to Earth is not enough. In fact, a safe round-trip without meaningful surface exploration would be regarded as a minimally successful space mission (Hoffman and Kaplan, 1997). Astronauts will need to be strong and healthy to explore the Martian surface and investigate the possibilities for human colonization. For this reason, the Hoffman and Kaplan report states that humans are the most valuable mission asset for Mars's exploration and cannot become the weakest link.

Any mission to Mars or another distant point beyond Earth orbit involving humans will likely be international in character and will likely include individuals from some combination of 13 countries, and because of current propulsion designs, the mission would be launched from the International Space Station or a similar orbiting platform. Crewmembers must arrive on Mars safely and in good health after a lengthy period of travel, probably 5 to 6 or more months (Figure 1-1).

After the journey to Mars, astronauts must be mentally and physically prepared to spend many months on the Martian surface in their spacecraft or ancillary habitation structure and to function productively for all of that time. Crewmembers must be trained and equipped to deal en route, on Mars, and during their return with medical emergencies such as an abscessed tooth, a fractured limb, or life-threatening conditions with no possibility for emergency evacuation. Spacecraft crews cannot expect to rely totally on the presence of physician-astronauts, who, if part of the crew, may themselves become sick or disabled. Finally, advice and support from mission controllers on Earth will be delayed, which is in contrast to today's almost instantaneous communication between mission controllers and space crews in Earth orbit. An astronaut's medical emergency could quickly turn into a tragedy if crewmembers were forced to depend on Mars-to-Earth round-trip audio transmission times of 20 to 40 minutes depending upon how far the spacecraft is from Earth.

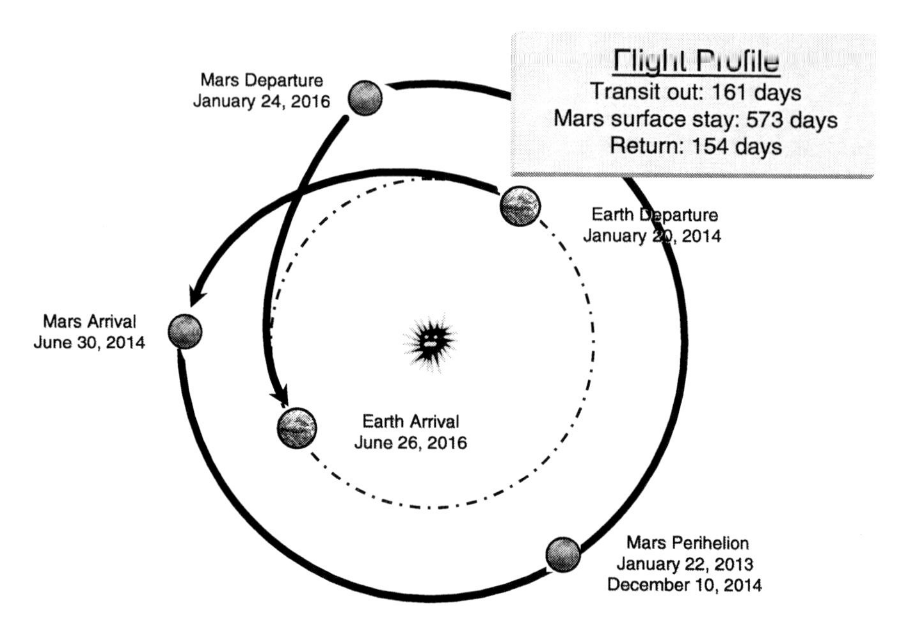

FIGURE 1-1 Trajectory of a human mission to Mars in 2014.
Source: Charles, 2000.

Despite such constraints, the medical goal on exploration-class space missions is to provide health care of sufficiently high quality so that crewmembers, once their exploratory mission is completed, can reasonably anticipate a safe return to Earth and, subsequently, healthy and productive terrestrial lives.

HEALTH RISKS OF SPACE TRAVEL

To date, some 350 persons have "flown" in space. Of these, most have flown for less than 30 days. Two cosmonauts, Vladimir Titov (366 days) and Valeri Poliakov (400 days), each spent more than a year in space, and U.S. astronaut Shannon Lucid traveled in space for 188 days. From this limited experience, it is clear that the health risks of prolonged exposure to microgravity and the confining environment of the spacecraft during space travel can be profound. Even 30 days in space can induce dramatic physiological changes. Some are minor and temporary, such as facial edema and an increase in height of up to an inch (Nicogossian et al., in press). Others are severe and may not always be reversible. For example, bone mineral density appears to decrease at an average rate of about 1 percent a month during exposure to microgravity, while recovery on Earth appears to proceed at a slower rate (Vico et al., 2000).

When the human skeleton is no longer exposed to gravity, a significant loss of bone mineral density occurs as a result of demineralization. The amount of bone demineralization varies; one Skylab astronaut lost nearly 8 percent of the mineral density in his calcaneus after 84 days in space, and a Russian cosmonaut aboard *Mir* reportedly lost up to 19 percent mineral density at the same site after 140 days (Vico et al., 2000). Thus, a Mars mission that exposes astronauts to microgravity for up to 3 years could theoretically result—if untreated—in a loss of bone mineral density of 50 percent or more at several structurally important skeletal sites. "Microgravity-induced bone loss—will it limit human space exploration?" This title of an editorial in *The Lancet* (Holick, 2000) asks a rhetorical question, for the loss of bone mineral density on long-duration space missions is one of the most serious and intractable health risks identified so far, and until this physiological effect of microgravity is resolved, a mission to Mars is unlikely to be undertaken with humans. So far, preventive interventions that NASA refers to as "countermeasures" have been only marginally effective ("countermeasure" is NASA's designation for preventive and therapeutic interventions before or during space missions) (Lane and Schoeller, 2000). If the loss of bone mineral density as a result of space travel cannot be surmounted by biomedical means, an engineering solution, for example, artificial gravity or some other means of integration of engineering and biology, will be necessary.

Radiation exposure is perhaps an even greater risk from travel beyond Earth orbit. Radiation beyond Earth's orbit is substantially different from the ionizing radiation to which humans are generally exposed on Earth because of the presence beyond Earth's orbit of high-energy charged solar and cosmic particles from deep space, ranging from protons to iron nuclei (SSB and NRC, 1996, 1998a,d, 2000). Little is presently known about the potential interaction of this nonionizing form of radiation with the DNA, cells, and tissues of astronauts (SSB and NRC, 1996, 1998a,d, 2000). There are no data on the effects of terrestrial exposure to such protons and high-atomic-number, high-energy particles that flood space beyond Earth orbit because of the current absence of experimental facilities on Earth. Furthermore, there is no way to predict when solar outbursts with their higher levels of radiation will occur (SSB and NRC, 1996, 1998a,d) and no current practical way to protect spacecraft crewmembers from them.

Until the radiation hazards to astronauts can be controlled or otherwise mitigated by physical shielding, a 1998 National Research Council report states, long-duration space travel should be postponed (SSB and NRC, 1998a). Even if an effective physical radiation shield is developed, it in no way diminishes the need for clinical study, including monitoring of crewmembers' exposures, long-term medical follow-up, and the development of preventive medical treatments to make astronauts more resistant to deep space-induced radiation damage. The anticipated cumulative time-dependent effects that both bone mineral density loss and radiation are anticipated to exert, furthermore, emphasize the need to develop alternative propulsion systems to decrease space travel time and the effects of microgravity and radiation from deep space.

Long-duration space travel can also be expected to pose risks of psychological and social stresses because of isolation, confinement, and living in cramped quarters for long periods of time. Without significant engineering improvements to the design of the spacecraft environment so that it is more biocompatible, there will likely be high noise levels, less than optimal light, and diminished privacy. In the view of some experts, these and other psychological stresses could turn out to be the most worrisome risk of all to astronaut health (SSB and NRC, 1998a, 2000). Habitability—or, more correctly, biocompatibility—thus must be an important initial consideration in spacecraft design.

Astronaut health requires a continuum of preventive, therapeutic, and rehabilitative care on the ground, during space travel, and upon the return from space travel. The continuum includes normal health maintenance and care for the physiological adaptations that humans experience as a result of the extreme environment of space (see Chapter 2). Furthermore, it must address a large variety of the minor and major medical problems, including psychiatric and behavioral health problems, and surgical problems that can develop among members of a group of individuals over extended periods of time in normal and extreme terrestrial environments (astronauts in training and between missions) and during extended periods in space beyond Earth orbit, which is the particular focus of this report (see Chapters 3, 4, and 5).

Although the preventive and rehabilitative aspects of health care are of utmost importance to maintaining a healthy, active astronaut corps, this chapter focuses on principles of health care during future long-duration space travel and habitation, for example, during exploration-class missions to Mars or colonization of Earth's moon. Other components of the continuum of astronaut health care before, during, and after space travel are covered in depth in the current report in chapters on behavioral health (Chapter 5), ethics (Chapter 6), and a comprehensive health care program for astronauts (Chapter 7).

Travel into deep space beyond Earth orbit involves many unique and hazardous elements:

- Isolation. Great distances preclude timely evacuation and a return to Earth for health care. Therefore, the crew must be prepared to deal in flight with diverse medical situations ranging from minor cuts to death.
- Limited resources. The spacecraft is unable to carry elements available on Earth because of storage space, power, and weight limitations.
- Closed environment. The spacecraft has a closed life-support system and cramped working and living quarters.
- Space-specific hazards. The space environment lacks gravity and contains damaging radiation.

Astronauts are a selected healthy population with low percentages of disease occurrence. However, it is still necessary to prepare for the worst case such as major trauma, appendicitis, obstructive cholelithiasis, acute myocardial infarction or disabling arrhythmia, stroke, pancreatitis, and death. Earth-based health problems will not be left behind and, thus, cannot be overlooked by relying on chance.

CHARGE TO THE COMMITTEE

With this as background, NASA in 1999 formally asked the Institute of Medicine (IOM) to "create a vision" for health care for astronauts traveling beyond Earth orbit. In a letter (see Appendix A) to IOM President Kenneth I. Shine, NASA Administrator Daniel S. Goldin pointed out that efforts to develop a more capable medical care delivery system in space had been internal to NASA. The focus had been on prevention, reflected in "strict" astronaut selection standards and close monitoring of their health status. So far that approach has succeeded, as astronauts have been free of major health problems during space missions, but spaceflights have been short and emergency evacuations from spacecraft in low Earth orbit have been possible. Evacuation and return to Earth will not be the case during future long-duration space missions beyond Earth orbit.

NASA requested IOM's assistance in "evaluating our current medical-care system and recommending the type of infrastructure we will need to develop to support long-duration missions, including interplanetary travel, in which timely evacuation of crewmembers will not be an option. Medical-care provider training, specialty mix, nonmedical crewmember skills, use of advanced technology, surgical/intensive care capability in space, rehabilitation approaches to cope with exposures to gravitational fields following exposures to microgravity, psychological/human factors challenges and use of robotics for health monitoring, education, and possible surgery are examples of the types of issues we would like you to address. We would also like you to consider the use of analog environments, such as remote Antarctic stations, for training and research. Ethical considerations in the face of limited medical-care capability are also important issues that need examination" (Goldin, 1999a). IOM asked experts in health sciences research and clinical medicine to address the health risks, medical needs, and patient care dilemmas that are likely to arise during long-duration space travel.

The committee's charge is shown in Box 1-1.

THE COMMITTEE'S STUDY OF HEALTH CARE FOR ASTRONAUTS TRAVELING BEYOND EARTH ORBIT

The committee has focused on the development of principles and general and developing practices for provision of the best possible health care to astronauts. This report covers the continuum of health care, from preventive services before departure, to treatment of conditions that might conceivably arise during long-duration space travel beyond Earth orbit, to health care on Mars and during the return to Earth. It also discusses the need for restorative and rehabilitative services for astronauts upon their return to Earth.

On the first exploration-class mission to Mars, expected to last nearly 3 years, the goal will be to keep the astronauts healthy, productive, and reasonably comfortable in an environment that is almost unimaginably distant and unforgiving. The IOM Committee on Creating a Vision for Space Medicine During Travel Beyond Earth Orbit envisions a health care system for astronauts that can deliver high-quality medical care, extensive psychological support, and excellent (albeit basic) surgical services to a special population of astronaut-patients who are unusually fit but also uniquely vulnerable. In the course of the committee's information gathering and data analysis five elements emerged as critical to the committee in addressing its charge.

- ***Risks to astronauts' health*** The committee sought to understand the

BOX 1-1
Charge to the Committee

Conduct an independent assessment of the current status of scientific knowledge, paying particular attention to pharmaceutical and technology principles, to provide optimal health care for astronauts during and upon return from spaceflights beyond Earth orbit.

Evaluate the most promising directions for the future of scientific progress in space medicine, which will include (1) identifying advances anticipated in related fields that might prove to be beneficial for medical practice in space and (2) exploring opportunities for finding innovative terrestrial medical care practices that have the potential to advance the theory and practice of space medicine.

Recommend a strategy at the national and international levels for medical care in space that holds the greatest promise for preventing and treating health conditions expected to develop during long-term spaceflight. The committee will provide recommendations to NASA regarding the most promising avenues for medical care compatible with the adaptability status of the crew. The committee will develop recommendations regarding the direction of future medical care investments to attract interest from scientists to this area of health science and medicine.

Suggest the most effective ways for NASA to address the priority areas in achieving this strategy. Assist in developing collaborative relationships between NASA and clinically prominent experts at the national level to advocate and guide the growth of space medicine, including the development of clinical practice guidelines as appropriate. Make recommendations on the distinctive contributions that could be made by the context of clinical areas supported by NASA, the federal agencies that perform health care (e.g., the U.S. Department of Veterans Affairs, the U.S. Department of Defense, and the Federal Aviation Administration), the pharmaceutical and biomedical device industries, and other organizations and agencies as appropriate.

Adopt and retain a flexible approach to this task, understanding the dynamic and changing nature of emerging knowledge in fields related to space medicine and in recognition of evolving needs of the space medicine community.

risks to human health during space travel, the extent to which astronauts are included in decision making about acceptable risks, and the extent to which society is informed regarding the risks and the possibility of a disaster.

- ***Astronauts as research subjects*** Certain data must be collected, analyzed, shared, and used to make conditions safer and better for those who follow. At the same time, what assurances do astronauts receive that any research in which they are required to participate is not only relevant but also essential? Are they informed of the risks and potential benefits?
- ***Clinical research*** High-quality health care for astronauts, as for any

individual, must be grounded in an evidence base, that is, documented and analyzed clinical observations supplemented by designed clinical research. The committee examined NASA's data collection and strategic plan for clinical research, the extent to which its research awards are open to the broadest possible scientific audience, the degree of hypothesis-driven research, and the degree of collaborative relationships. To what extent is the data-gathering effort ongoing, rigorous, prospective, and methodologically sound? The committee examined the potential of the International Space Station as an orbiting clinical research laboratory to focus on better understanding the effects of microgravity on humans.

- ***Astronaut health care*** The committee evaluated the general continuum of health care that includes premission, intramission, and postmission preventive medical and dental care and health education as well as the traditional forms of medical, surgical, and behavioral medical care focused on the unique environment of extended periods in an isolated and remote, self-contained "capsule."
- ***Crew selection*** Selection of the crew for the first mission beyond Earth orbit will be critically important. The committee focused its attention on group and individual characteristics, selection, and training.

These are a few of the areas the committee sought to understand in fulfilling its charge to provide a vision for space medicine during travel beyond Earth orbit.

In going about its work, the IOM committee held five information-gathering meetings over the course of 15 months (October 1999 through December 2000), with one additional closed meeting held in January 2001 for discussion of the committee's conclusions and recommendations. At least a portion of each of the first five meetings was open to the public. The committee's first meeting featured presentations by officials from NASA headquarters in Washington, D.C., and the Johnson Space Center in Houston, Texas (see Appendix A). The committee's second meeting, in February 2000, incorporated a site visit to the Johnson Space Center and included briefings by some 30 NASA clinicians and researchers in space medicine (see Appendix A). Two of the committee's meetings included workshops open to the public. The first workshop, held in Washington, D.C., in April 2000, focused on the dental health needs of astronauts on long-duration space missions (see Appendix A). The second workshop, held in Woods Hole, Massachusetts, in July 2000, focused on what has been learned about maintaining the health and well-being of scientists and explorers on duty in remote and extreme environments for long periods of time (see Appendix A). The July workshop also included a panel discussion and a question-and-answer session with five physician-astronauts.

This report extends the findings and recommendations of earlier National Research Council reports that are focused on a variety of basic biomedical concerns and that were developed at the request of NASA and published in 1987, 1996, 1998, and 2000 (SSB and NRC, 1987, 1996, 1998a, 2000). It extends the findings and recommendations to the issues that directly influence the well-being and delivery of health care to astronauts during space travel beyond Earth orbit. The current committee and its report differ substantially from previous National Research Council committees and reports in that the committee focused on human clinical research and astronaut health care. Each earlier report assessed the then-current state of health sciences research that appeared to be of importance to the future development of space medicine. The reports laid out long-term strategies for future biomedical research. The current report goes to the next level by focusing on principles that can be used to build a health care system that ensures the health and safety of humans during long-duration missions beyond Earth orbit. The earlier reports also cautioned, as this one does, that substantial reductions in anticipated risks must be demonstrated before humans are sent on missions beyond Earth orbit. Professionals in engineering and biomedicine must work together to make missions beyond Earth orbit succeed. In extending that caution, this committee report suggests principles and directions for space medicine (see Box 1-2 for the definition of space medicine that the committee used as its reference) that NASA should consider in its role as a leader in developing this new frontier to provide the optimal health care for individuals who will be embarking on and returning from missions beyond Earth orbit.

BOX 1-2
What Is Space Medicine in Reference to Developing a Vision and Strategy for Astronaut Health?

For the purpose of the present study the committee has considered *space medicine* to be a developing area of health care that has roots in aerospace medicine but that is focused on the health of individuals so that they can perform in and return in good health from increasingly distant extreme space environments, for example, from short-duration space capsule flights, space shuttle flights, missions to the Moon, long-term Earth-orbiting space station missions, and in the next stages, exploration-class missions beyond Earth orbit, including missions involving planetary colonization.

This view of space medicine is consistent with NASA's, in which *operational space medicine* is defined as "a distinct discipline of medicine focused on the unique challenges of human spaceflight and the medical risk management of associated hazards" (from the Mission Statement of the Medical Operations Branch, Johnson Space Center, 2000).

Presently, the specialty of aerospace medicine includes aviation medicine and space medicine. There are approved residency programs as well as recognition by all regulatory agencies and the American Medical Association. The Aerospace Medical Association is the established professional home providing scientific meetings as well as a monthly specialty journal (*Aviation Space Environmental Medicine*) for all practitioners and scientists whether in aviation or space medicine.

The understanding and definition of space medicine will change as space travel and medicine change. Those changes must continually be kept in mind as visions and strategies for space medicine and astronaut health and safety evolve.

NOTES

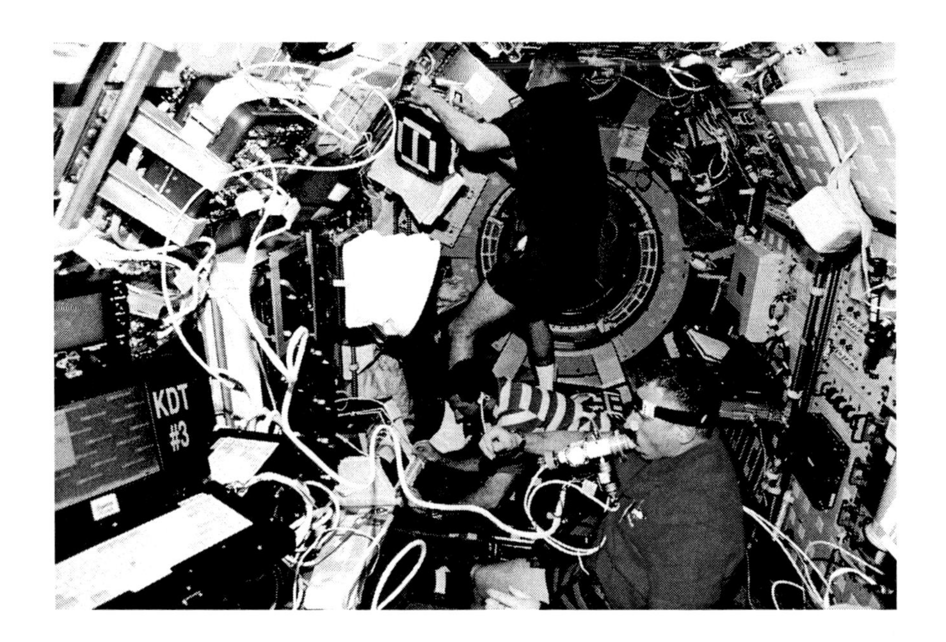

Payload Specialist Jay C. Buckley, Jr., Payload Commander Richard M. Linnehan, and Astronaut Dafydd R. (Dave) Williams (left to right) during pulmonary function tests in support of the Neurolab mission aboard the Earth-orbiting space shuttle *Columbia*, April 24, 1998, during the STS-90 mission. NASA image.

2
Risks to Astronaut Health During Space Travel

. . . we must assume that for a long time to come (although not forever), weightlessness will be an obligatory condition of space flight. For this reason, all aspects of this issue must be considered from the point of view of the possibility of functioning in microgravity.

G. I. Meleshko, Y. Y. Shepelev, M.M. Averner, and T. Volk, 1994

OVERVIEW

Over the life of the U.S. space program, generations of astronauts have learned how to live and work in weightlessness. Humans evolved in gravity; how the body would function in its absence or near absence was an unanswered question. Would it be possible to eat and drink in microgravity? Would it be possible to perform complex tasks? The early answers were affirmative. Thus, selected fit and healthy humans have been sent into space for three decades and have functioned well (Lane and Schoeller, 2000).

Although humans have adapted to weightlessness, readapting to Earth's gravity is problematic. Exposure to microgravity affects the body in many ways. Some effects are severe and long lasting, such as loss of bone mineral density. Others are minor and temporary, such as facial puffiness due to fluid shifts (Nicogossian et al., in press). It is unlikely that all effects of microgravity are known, and surprises may yet be in store as humans venture longer and farther into space. This chapter is about human physiological adaptation to space travel. In it, the Institute of Medicine (IOM) Committee on Creating a Vision for Space Medicine During Travel Beyond Earth Orbit examines what is known about the effects of microgravity and space travel on the human body. This is the starting point for generating priorities in clinical research and health care for space travel beyond Earth orbit.

In developing this chapter, the committee has relied heavily on briefings and published information from the study's sponsor, published scientific articles, and two recent National Research Council (NRC) reports. The first report, *A Strategy for Research in Space Biology and Medicine in the New Century* (SSB and NRC, 1998a), provided a science-based assessment of the most important biomedical research topics in 1998 to be pursued over the next decade. The second report, *Review of NASA's Biomedical Research Program* (SSB and NRC, 2000), examined the National Aeronautics and Space Administration's (NASA's) biomedical research enterprise 2 years later and measured it against the plan set forth in the earlier report. The IOM committee endorses the findings and recommendations of both NRC reports.

The current report extends the vision of the two previous NRC reports to clinical research and clinical care in space. This chapter responds to the portion of NASA's charge to the IOM committee to "conduct an independent assessment of the current status of scientific knowledge" relevant to providing optimal health care for astronauts traveling beyond Earth orbit. In so doing, the chapter describes the effects of weightlessness and space travel on the physiology and functioning of the human body. It discusses the evidence on which the findings are based, the steps that need to be taken, and the research challenges and opportunities that lie ahead.

Most of what is known about the effects of microgravity on the human body has been learned on short missions into space. NASA is now looking ahead to longer-duration space missions, initially in Earth orbit and later into deep space. Over the next decade, a number of astronauts will have 3- to 6-month tours of duty aboard the International Space Station (ISS). These may be followed by extended stays on the Moon or exploration-class missions to Mars, or both. Before the United States and its international space partners commit to any such plans, however, there needs to be a better and fuller understanding of the risks to astronaut well-being and the safety of long-duration space travel in and beyond Earth orbit.

The chapter presents numerous examples of the effects of exposure to microgravity and space travel on human physiology (Box 2-1). The examples

BOX 2-1
Some Major Human Physiological Changes Resulting from Extended Travel in Earth Orbit

Musculoskeletal System
- Loss of bone mineral density
- Loss of skeletal muscle

Cardiovascular System
- Orthostatic hypotension
- Loss of hydrostatic pressure

Pulmonary System
- Changes in pulmonary circulation and gas exchange

Alimentary System
- Ileus
- Decrease in absorption or malabsorption

Nervous System
- Ataxia
- Motion sickness
- Disturbed fine motor and gross motor functions
- Altered sleep-circadian rhythm and sleep deprivation

Reproductive System
- Effects of radiation on gametes

Urinary System
- Renal calculi

Hematological and Immunological Systems
- Anemia
- Potential immunologic depression

Source: Billica, 2000.

are by no means exhaustive, however. The material in this chapter is arranged by organ system, with those for which the physiological effects are best documented presented first. The chapter also includes a discussion of future methods for the monitoring of astronauts' health status—an important aspect of detecting, understanding, and countering the untoward physiological changes that may affect astronaut well-being and mission performance.

Finally, the chapter includes a discussion of the comprehensive, long range approach to clinical research that NASA needs to consider implementing to best protect human health and safety during long-duration space travel. Historically, NASA has faced difficulty in conducting clinical research in space medicine. One problem is the small numbers of research subjects (astronauts) available for study. The overriding reason, however, is that microgravity cannot be duplicated on Earth; it can only be approximated. The terrestrial means of research on bone mineral density loss in microgravity are bed rest, immersion in water, or immobilization. All have their own disadvantages. The opening era of the permanent presence of humans in Earth orbit on the ISS in October 2000, however, provides an enduring test bed that will eventually help provide an understanding of human physiology in microgravity.

Countermeasures to Solve Physiological Adaptations to Space

Faced with the necessity to maintain astronauts' health during periods of exposure to microgravity and other extreme conditions of spaceflight, NASA has pursued the development of preventive and counteracting measures (i.e., countermeasures) to guard against or reverse the potential pathophysiological effects of space travel. A variety of countermeasures have been used in longer-duration spaceflights (Mikhailov et al., 1984; Bungo et al., 1985; Greenleaf et al., 1989; Fortney, 1991; Arbeille et al., 1992; Cavanagh et al., 1992; Charles and Lathers, 1994; Hargens, 1994; Convertino, 1996b). The American and Russian space programs use different strategies. Some examples of countermeasures that had been developed as of 2000 include subcutaneous injections of erythropoietin to prevent decreases in erythrocyte mass and vigorous in-flight exercise regimens to reduce loss of bone mineral density. So far, countermeasures appear to be largely ineffective, but the data are sparse (Bungo et al., 1985; Buckey et al., 1996b; Convertino et al., 1997; Lane and Schoeller, 2000).

NASA's general approach to the development of countermeasures was presented to the IOM committee at the Johnson Space Center (Paloski, 2000; Sawin, 2000). The rationale (Figure 2-1) outlined a number of steps that have been incorporated into NASA's Countermeasure Evaluation and Validation Project, which can be summarized as follows:

1. conduct research to understand the basic nature of the physiological problem,

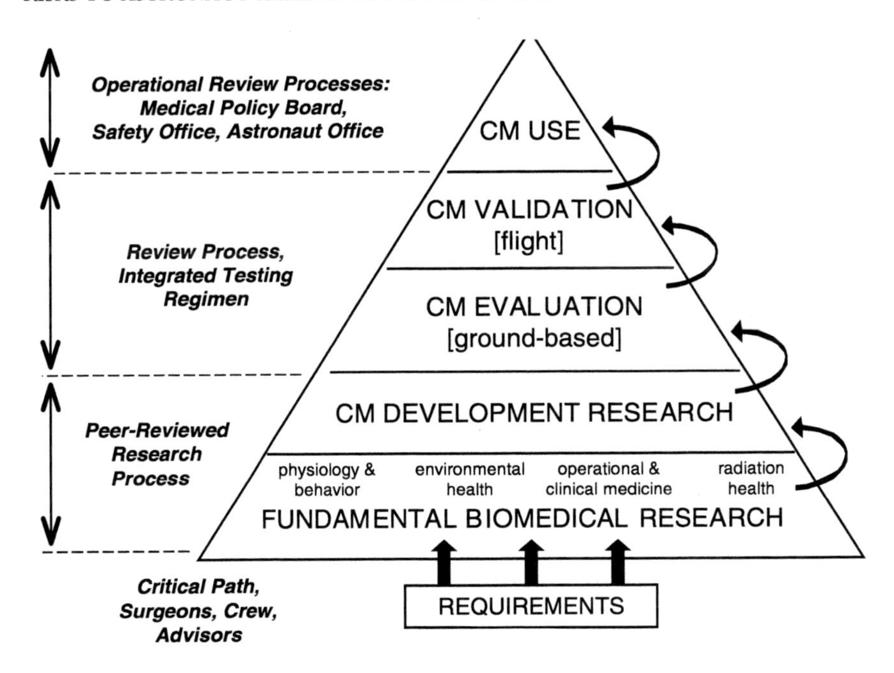

FIGURE 2-1 Countermeasure (CM) evolution. Source: Paloski, 2000.

2. formulate a countermeasure strategy based upon that physiological understanding,
3. test the countermeasure and demonstrate its efficacy on the ground, and
4. validate the countermeasure in space.

It has been difficult for NASA to design and test effective countermeasures, and no single countermeasure has yet to be validated as clinically efficacious. The potential for better design and evaluation of countermeasures improved dramatically on October 30, 2000, with the arrival of the first multinational astronaut crew to inhabit space as residents on the ISS. The ISS offers NASA and its international partners a longer-term orbiting clinical research laboratory in microgravity to investigate—and ultimately prevent—the adverse changes in human physiology described in the pages that follow. To assume that a terrestrial model duplicates the physiological effects of microgravity is a logical flaw that could lead to reliance on ineffective countermeasures. The plan outlined by NASA therefore demands that a significant amount of physiological research be conducted on the ISS and immediately following long-duration space missions on the ISS. The amount of research required, the duration of the research, and the extensive nature of the research will have to be considered in the planning phases of ISS missions.

This chapter also discusses new and future methods for the diagnosis and monitoring of astronaut health status in space and NASA's health risk assessment and management process that includes countermeasure development. The chapter concludes with a discussion of the comprehensive long-range approach to clinical research that NASA needs to consider to prepare for successful long-duration missions with humans beyond Earth orbit.

MUSCULOSKELETAL SYSTEM

Loss of Bone Mineral Density in Microgravity

Changes in bone mineral density, muscle mass, and muscle function are the best-documented physiological effects of human space travel. The loss of bone mineral density in microgravity is well documented (Vico et al., 2000). Serious acute consequences of bone mineral density loss (i.e., fracture and the formation of renal stones) as well as long-term morbidity may complicate long-duration space travel beyond Earth orbit. Working in microgravity within a spacecraft, during extravehicular activity, and upon a low-gravity moon or planet presents many increased risks for bone fracture and the necessity for wound healing. From a practical viewpoint, virtually nothing is known about how microgravity will affect fracture management and healing during long-duration space missions. For example, is it better to cast, internally fixate, externally fixate, or electrically stimulate a fracture sustained on Mars? The committee was unable to locate data from studies with animals or humans or from basic, translational, or clinical studies on these clinical treatments issues; but knowledge about such clinical treatments issues will be important to sustain human health and performance should a bone fracture occur during space travel beyond Earth orbit.

At the basic science level, little is known about the fundamental mechanisms underlying the loss of bone mineral density in microgravity; hence, scant progress has been made on the development of effective countermeasures. This must be an extremely high priority before long-duration space travel can be deemed reasonably safe with regard to the risk of fractures, the associated increased risk of renal stones, and basic skeletal support.

Small numbers of subjects and, in many cases, incomplete data have hindered clinical studies, resulting in a lack of reliable databases (SSB and

NRC 1998b; Lane et al., 1999; Lane and Schoeller, 2000). Thus, it is difficult, if not impossible, to generate reliable conclusions that can be applied to individual astronauts. A recent review (Smith et al., 1999) focusing on calcium metabolism after a 3-month space mission showed an approximately 50 percent increase in the level of calcium absorption accompanied by a 50 percent increase in the levels of both calcium excretion and bone resorption, as determined by calcium kinetics and bone marker analyses, respectively. Subjects lost approximately 250 milligrams of bone calcium per day during space travel in Earth orbit and appeared to regain it at a slower rate after their return to Earth. There is a suggestion from limited studies of bone density markers in serum and urine, which are used to approximate the relative rates of bone formation in relation to rates of bone resorption, that an uncoupling of the two processes led to an imbalance, with bone resorption predominating (Caillot-Augusseau et al., 1998). Again, because of the small numbers of subjects, this conclusion cannot be generalized and may not be applicable to all astronauts.

As shown in Table 2-1, mainly weight-bearing bones (spine, neck, femur, trochanter, and pelvis) lost bone mineral density during space missions in Earth orbit: on average, greater than 1 percent per month for cosmonauts on the Russian *Mir* space station. In contrast, there was no significant loss from bones in the upper extremity (arm).

Additional measurements ($n = 40$) that include shorter-term (<3 weeks) U.S. space shuttle flights show that loss of bone mineral density begins within a few days and continues for the longest period measured (1 year) without showing signs of leveling off. It is noteworthy that the standard deviations in all studies are high, which may indicate wide variations in the responses of individuals. This suggests that there may be substantial phenotypic (and presumably genotypic) variations in susceptibility to microgravity-induced bone mineral density loss. If shown to be true, this concept would have important implications for the selection of crewmembers for long-duration missions (e.g., are menopausal female astronauts or specific male or female astronauts phenotypically at greatly enhanced risk?).

The collection and analysis of clinical data for a comprehensive database to determine whether individuals with phenotypes that make them "resistant" to and "at risk" for space travel-induced bone mineral density loss exist represent but one set of important challenges for NASA that has been identified. Understanding these patterns in relation to basic patterns of up- and downregulation of gene arrays as a result of exposure to microgravity could lead to the development of specific interventions to prevent microgravity-induced bone mineral density loss.

TABLE 2-1 Average Bone Mineral Density Loss on *Mir*

Variable	Number of Crewmembers	Mean Loss (Percent/Month)	Standard Deviation
Spine	18	1.07[a]	0.63
Neck of femur	18	1.16[a]	0.85
Trochanter	18	1.58[a]	0.98
Total body	17	0.35[a]	0.25
Pelvis	17	1.35[a]	0.54
Arm	17	0.04[a]	0.88
Leg	16	0.34	0.33

[a] $p < 0.01$.
SOURCE: LeBlanc et al., 1996.

Associated with the bone mineral density loss is a rather consistent hypercalciuria, which increases the risk of formation of renal stones (Schneider et al., 1994). Importantly, this limits pharmacological options such as treatment with normal dietary supplements (e.g., calcium and vitamin D).

Reversibility, Genetic Variability, and Mechanism of Bone Mineral Density Loss

Although data are limited, it appears that changes in calcium metabolism and bone mineral density are reversible. There are suggestions, however, that reversal of the changes is slower than their evolution and that the rate and extent of reversal are highly variable (Vico et al., 2000). A mission to Mars, for example, would involve a period of low or nearly zero gravity during space travel, a second period of time in gravity well lower than that on Earth while the expedition was on the surface of Mars, a third period of time again spent in low gravity during the return flight, and the ultimate return to Earth's gravity. How these sequential changes in gravity loading will influence bone mineral density loss is unknown. It is possible that they may exacerbate and accelerate the incidence of Earth-based diseases such as osteoporosis. As such, it may be appropriate to consider terrestrial referenced 6- and 9-year windows of disease incidence instead of the disease incidence for a 3-year period.

Limitations in data collection and analysis, the small sizes of databases, the lack of precise bone mineral density measurements (which have a 1 to 2 percent coefficient of variation), and the very high natural variations of markers of bone mineral density turnover all contribute to the difficulty of obtaining reliable data that would be useful for clinical decision making for space travel. Therefore, the most reliable and efficient clinical studies must probably use individual astronauts as their own controls. The etiology of bone mineral density loss is multifactorial and likely polygenic, like most common diseases. The basis for or the implications of individual variability is not known, further complicating interpretation of the limited clinical data. Long-duration space missions in Earth orbit offer the opportunity to obtain crucial data by careful clinical research. These data could then be used to generate hypotheses and to guide measures to protect subsequent astronauts from unnecessary morbidity and even death.

Additional factors that may affect the relative risk of bone mineral density loss during space missions are bone mass as measured by bone densitometry and bone turnover rates as measured by markers of bone density in serum and urine determined before the initiation of a space mission. Although the selection of astronauts with higher bone masses may not prevent bone mineral density loss, it may prevent the consequences of decreased bone mineral density; that is, bones with higher mineral densities may be at decreased risk of fracture. Thus, because of unknown genetic components and the variability among astronauts with regard to the rate of bone mineral density loss, it might be increasingly important to identify individuals whose bodies are able to resist bone mineral density loss on prolonged space missions (Vico et al., 2000). This can be accomplished only by extensive premission, intramission, and postmission analyses of bone mineral density and markers of bone mineral density turnover in serum. It may be possible to identify individuals who are less prone to the effects of microgravity on bone mineral metabolism and bone mineral density loss. This knowledge not only might have implications for space travel, but it also may well provide important information about diseases such as osteoporosis.

A number of animal models, including models of unloading, have been used to simulate weightlessness (SSB and NRC, 1998a; Lane and Schoeller, 2000). It is unknown whether such models reflect the physiology of humans during space travel. Rats have been used in microgravity experiments, but the results of those experiments may be of limited value given the poor correlation between rat bone metabolism and the bone architecture in humans (SSB and NRC, 1998a, 2000). The development of land-based animal models for investigation of the pathophysiology and pharmacotherapeutics of bone mineral density loss during space missions beyond Earth orbit will require more knowledge of bone mineral density loss in humans. Although the advantage of using experimental and natural animal models, for example, hibernating bears (Harlow et al., 2001), is evident—in that it allows more focused investigation into basic molecular and cellular mechanisms—successes from the study or development of such models will likely lag behind the need to test the effectiveness of therapeutic interventions in humans during space missions beyond Earth orbit. The use of immobilized human subjects on Earth as models may have its place. However, the high priority for the evaluation of the effectiveness of countermeasures in the microgravity environment renders such models less valuable in the short term.

The technology for the accurate testing of bone mineral density is improving and is becoming increasingly miniaturized. It will probably become necessary to assess bone mineral density changes during space missions both in and beyond Earth orbit with reasonable precision for clinical research purposes. It is anticipated that this technology will be useful in assessments of the need for and testing of the effectiveness of specific interventions during space travel. In addition, it is anticipated that better documentation of the validity of bone mineral density markers in evaluations of the mechanisms of bone mineral density loss would also provide a means for optimization of highly targeted interventions. Therefore, the development of a means to measure such markers in the long-duration space mission environment should be a high priority in conjunction with an aggressive NASA countermeasure research program. Better measures of bone integrity, for example, three-dimensional assessments, not just bone mineral density, are also needed. Ultimately, intramission tests should be the indicators that guide specific therapies.

Clinical Research Opportunities in Astronaut Physiology and Health

Clinical Research Opportunity 1. Establishing the course of changes in bone mineral density and markers of bone mineral density turnover in serum and urine before, during, and after space travel.

Clinical Research Opportunity 2. Developing a capacity for real-time measurement of bone mineral density and enhanced three-dimensional technology to assess the risk of fracture during space travel.

Because of the significant current research effort into the prevention and treatment of osteoporosis, new techniques that better assess bone mineral density turnover and fracture susceptibility can be anticipated within the next decade. These techniques should be evaluated, adapted if necessary, and incorporated into NASA's clinical research program.

Clinical Research Opportunity 3. Identifying human phenotypes and genotypes resistant to space travel-induced bone mineral density loss.

This may involve analysis of both data from large phenotypic databases and genetic factors as they become available. Recent data suggest a wide spectrum in both the level of acute bone mineral density loss and return to normal bone density upon the return to Earth (Vico et al., 2000).

Clinical Research Opportunity 4. Tailoring therapeutic interventions (i.e., countermeasures such as diet, exercise, and medications) as a high priority and then validating the promising countermeasures in studies with astronauts during exposure to microgravity.

Pharmacological countermeasures deserve special emphasis. They may need to be tailored to an individual's response, and ongoing monitoring of effectiveness may be required during long-duration space travel beyond Earth orbit. For most of the clinical studies, astronauts must be their own controls, which again requires the ability to conduct comprehensive analyses of bone-related metabolic parameters during space travel.

Effects of Microgravity on Skeletal Muscle

During space travel, the primary effects of microgravity on skeletal muscles include the deterioration caused by the lack of gravity on slow-twitch muscles, with conversion from slow to fast muscle fiber type; and decreased fiber size in rats (Riley et al., 1990; Jennings and Bagian, 1996). Young rats subjected to 18 days of hind-limb unloading developed an abnormal gait that persisted, suggesting permanent damage to the neuromuscular pathway (Walton et al., 1997). Significant atrophy has been observed in human muscle after only 5 days in space, but the time course of deterioration has not been established. Moreover, it is not known when or whether a plateau might be reached. In addition, the foot-drop posture in microgravity shortens the extensor compartment and appears to accelerate loss of the thick filament (Edgerton et al., 1995).

Most of these primary changes to skeletal muscle during space travel appear to represent simple deconditioning without apparent pathology and can be considered appropriate adaptations for functioning in a low-workload microgravity environment. However, an abrupt return to gravity imposes high workloads on weakened muscles and post-space travel pathologies: muscle fatigue, weakness, incoordination, and delayed-onset muscle soreness. In rats, changes in microcirculation occur during space travel as a result of the cephalad fluid shift, and reloading-induced edema and ischemic tissue necrosis may occur upon the return to Earth's gravity (Riley et al., 1996). In addition, adaptation to the lower workload in microgravity may render muscle tissue more prone to structural failure when it is reloaded.

A variety of exercise-based protocols have been used, but none have been adequately validated and none have proved to be more than modestly effective. Cycle ergometry has successfully been used to ameliorate cardiovascular aerobic deconditioning but did not prevent muscle deterioration. However, bicycle ergometry and treadmill exercises counteract the tendency to attain the foot-drop posture in microgravity that shortens the extensor range and that appears to contribute to accelerated loss of the thick filament (Edgerton et al., 1995; Widrick and Fitts, 1997).

The *Task Force on Countermeasures: Final Report* (NASA, 1997) concluded that existing cycling, rowing, and treadmill exercise protocols did not maintain muscle mass or a positive nitrogen balance. A significant contributing factor to this loss of muscle protein may be inadequate nutrition during space travel. To correct this, lower-body negative pressure coupled with resistance exercises should be tested as a means of maintaining the microcirculation and muscle strength (Koslovskaya et al., 1990). NASA funded a lower-body negative-pressure investigation in 1999 to study a countermeasure that uses developments in advanced technology. In addition, the possibility of pharmacological intervention apparently has not been explored, and better monitoring and recording of the history of muscle use and condition (e.g., specific exercise and nutrition history) during space travel need to be established to reduce uncontrolled parameters.

CIRCULATORY AND PULMONARY SYSTEMS

Orthostatic Hypotension

The single most significant problem associated with the cardiovascular and pulmonary systems as a consequence of microgravity appears to be orthostatic hypotension, a fall in blood pressure and the associated dizziness, syncope, and blurred vision that can occur when one stands up or simply stands motionless in one position. In the context of space travel, a major cause of orthostatic hypotension is the persistent lowering of peripheral vascular resistance during space travel, and more than two-thirds of all astronauts experience orthostatic hypotension when they reenter Earth's gravity (Buckey et al., 1996a,b). In other words, orthostatic hypotension is a physiological adaptation to microgravity that readapts after exposure to gravity. Resetting of the baroreceptor reflex occurs over a period of several days, depending upon the individual, the duration of space travel, and the individual's fluid status (Fritsch-Yelle et al., 1994). Symptoms can be uncomfortable or annoying, or both, but life-threatening complications have not been reported.

It is generally accepted that there is at least a general relationship between the duration of exposure to microgravity and the duration of the period of recovery from the orthostatic hypotension. However, few data are available concerning the severity of orthostatic hypotension, and the available information concerning the relationship between the duration of exposure to microgravity and the degree and duration of orthostatic hypotension is inadequate. No information is yet available on how orthostatic hypotension might be affected by the 0.4G gravity on Mars compared with how it is affected by the 1G gravity on Earth. It seems reasonable to assume, however, that on an exploration-class mission to Mars, orthostatic hypotension would represent a major problem if emergency procedures were to require the crew to move about immediately after landing.

Clinical Research Opportunity 5. Considering artificial gravity and pharmacological interventions as solutions to orthostatic hypotension.

Other Effects on the Cardiovascular System

The cardiovascular system undergoes several other physiological adaptations to microgravity. During weightlessness, there is a loss of hydrostatic pressure, especially in the lower extremities. Fluid shifts from extravascular to intravascular spaces and toward the upper part of the body (Thornton et al., 1987; Leach et al, 1996). This provokes objective and subjective symptoms, especially in the first days of space travel. The baroreceptors in the carotid arch sense a relative central hypervolemia and induce neurohormonal mechanisms that lead to diuresis and hypovolemia. Central venous pressure drops from the normal range of 7 to 10 mm Hg to 0 to 2 mm Hg (Kirsch et al., 1984; Buckey et al., 1996a).

The prolonged reduction in central venous pressure during exposure to microgravity resets the baroreceptors to a lower operating point; this in turn limits plasma volume expansion during attempts to increase fluid intake. Ground-based and in-flight experiments have demonstrated that restoration and maintenance of plasma volume before the return to Earth's gravity and an upright posture may return the central venous pressure to normal levels and reset the baroreceptor reflex (Convertino, 1996a). Microgravity also produces a decrease in renal and femoral vascular resistance, with maintenance of cerebral flow at rest. During orthostatic testing, lower-limb vascular resistance does not compensate for the fluid shift (Arbeille et al., 1996).

The changes in intravascular volume during space travel lead to corresponding changes in stroke volume and cardiac output. For example, corresponding decreases in stroke volume and cardiac output have been observed over a range of atrial pressures. Echocardiographic experiments performed during space shuttle flights showed an average decrease in stroke volume of 15 percent, measured after a period of 3 days of adaptation to weightlessness (Pantalos et al., 1998). The heart rate is generally unchanged. Thus, although the cardiovascular system demonstrates many changes during space travel, to date few have appeared to be severe enough to affect human health or performance during space travel. NASA has, however, identified potentially serious cardiac dysrhythmias, impaired cardiovascular responses to orthostatic stress, diminished cardiac function, manifestation of previously asymptomatic cardiovascular disease, and impaired cardiovascular responses to exercise (Levine et al., 1996) as potentially serious risks during space travel (see Table 2-2 later in this chapter).

These alterations in cardiac function could also present serious risks to astronaut health during long-duration space travel. However, the incidence and the severity of the dysfunction in each of these categories during short- and long-duration space missions are unknown. Currently available data (SSB and NRC, 1998c, 2000) have not been critically peer reviewed and do not convince the committee that cardiovascular risks should be labeled "serious" at this time. Substantial new research is required to define the degree of risk, the likelihood and severity of cardiac alteration, and reasonable therapeutic options for long-duration space travel.

Effects of Microgravity on the Pulmonary System

Thus far, no significant problems have been identified on the basis of the observed changes in pulmonary physiology in microgravity. Reported observations include a change in the pattern of ventilation in microgravity that results in decreased tidal volume and an increased frequency of ventilation (Prisk et al., 1995b, West et al., 1997). There is decreased dead space with normal oxygen uptake and an improved carbon dioxide diffusion capacity (Prisk et al., 1995a,b). The overall improvement in lung function cannot be completely explained by what had been supposed to be gravity-induced ventilation/perfusion ratio inequalities, raising fundamental questions about normal gas exchange physiology (Prisk et al., 1993, 1995a; Lauzon et al., 1998). Two potential risks to pulmonary function are dysbarism and a patent foramen ovale, although the latter is really a potential end risk to the nervous system. NASA is aware of these potential risks to astronaut health and is actively investigating what measures, if any, are needed to decrease the risks.

ALIMENTARY SYSTEM

Nutrition in Space

NASA has conducted extensive research on nutrition issues pertaining to spaceflight and has conducted such studies in collaboration with academic specialists (Lane and Schoeller, 2000). Despite 35 years of nutrition research, however, weight loss, dehydration, and reduced appetite continue to challenge NASA food scientists (Lane and Schoeller, 2000).

Space Motion Sickness

One of the most common conditions requiring pharmacological treatment in a microgravity environment is space motion sickness (SMS) (Putcha et al., 1999). SMS is a syndrome consisting of headache, malaise, disorientation, nausea, or vomiting. Putcha and colleagues (1999) reported that 47 percent of unit doses of medication given during space missions were for the treatment of SMS. These usually consisted of promethazine with or without dextroamphetamine, although the frequency of drug administration and the routes of administration were not reported.

The etiology of SMS is not known, although fluid redistribution (Simanonok and Charles, 1994) and alterations in bowel motility (Harris et al., 1997) are factors that have been implicated. Although SMS is largely considered a "minor" disorder, its effects are serious enough that extravehicular activity is prohibited during the first 3 days of a mission to avoid the possibility of vomiting and aspiration into the space suit (Simanonok and Charles, 1994). Given the high frequency of occurrence of SMS, its ability to affect an astronaut's function, and the high frequency of use of medication with potential adverse effects that affect function, SMS is an area for clinical investigation on longer ISS missions.

Many astronauts who develop symptoms of SMS also seem to develop a transient gastrointestinal ileus, diagnosed by an absence of bowel sounds. Although motility may remain decreased throughout flight and bacterial populations may change, data from short-term spaceflights do not suggest that this leads to significant medical problems (Lane and Schoeller, 2000). The etiology is unclear. Microgravity-induced movement of the abdominal contents within the abdominal cavity, simulating a surgical ileus, may be responsible. Another possibility is neurohumoral mediation.

Individuals who lack bowel sounds during the first 48 hours in space and who attempt to ingest food will often vomit, just as postoperative patients do. The simplest and most effective treatment appears to be patience. Those individuals who wait until they adjust to zero gravity and develop audible bowel sounds before they ingest solid food seem to do well. Most crewmembers on spaceflights resolve the ileus spontaneously within 48 hours, and adequate hydration should be maintained during that time, particularly if there is no intake of solid food. This gastrointestinal problem appears to be self-limiting, and overtreatment with laxatives may cause diarrhea, which can be a difficult problem to deal with in the space environment. No additional predictable problems with ileus would be expected on long-term spaceflights. Valuable information may be gathered from astronauts with extended stays on the ISS, and further clinical data may be obtained from studies investigating the etiology of SMS.

Clinical Research Opportunity 6. Investigating the relationship between space motion sickness and absent bowel sounds including pharmacological and adaptive countermeasures.

NERVOUS SYSTEM

Useful data on neurovestibular function, sleep (circadian rhythms), eye-hand coordination, fine and gross motor functions, and visual perception and reorientation have been collected in real-time, post-space travel studies and from studies with animals. However, little or no readily accessible data are available on sensation and proprioception. Some data are available, however, from isolated and confining analog situations describing predictors of behavior during long-duration missions (Palinkas et al., 1998).

Neurovestibular Function

Several studies have been performed to assess the effects of microgravity, nonvertical positioning, and simulated gravitational environments (short-arm centrifuge) on the neurovestibular system. These studies have focused on the adaptation to—and the relationship between—the visual system and body position during travel in space (Cohen et al., 1995). Countermeasures have been designed to test the effects of the centrifuge apparatus or body tilt on postural stability and possibly orthostatic tolerance (Black et al., 1999). Disconjugate gaze occurs and persists for weeks after space travel (Markham and Diamond, 1999). Markham and Diamond discuss the use of medication (promethazine) for SMS, including its impact on alertness, the mechanism of delivery, and possible antidotes (amphetamines) for drowsiness. No results of controlled trials, dose-response trials, or comparative efficacy studies have been published. Recent NASA Neurolab studies, however, deemphasize the potential importance of overt neurovestibular disturbance during space travel.

Sleep and Circadian Rhythm

Evidence is accumulating that sleep is disrupted during space travel and that the circadian rhythm is disrupted (Box 2-2). This may be mediated through the neurovestibular system. Data from *Mir* suggest that the duration of sleep is reduced, that sleep is not as deep, and that in other ways sleep may be physiologically different from the sleep experienced on Earth (Putcha et al., 1999). Possible explanations include the low levels of light in the spacecraft, changes in light-dark cycles, and the misalignment of work-rest shifts with light cycles.

Long-duration space travel may produce even more aberrant sleep disruptions and associated abnormalities. These can result from disruption of the hypothalamic-pituitary axis, with the resultant release of growth hormone, and changes in cortisol peaks and valleys. Disruptions of the circadian rhythm may result in abnormal stress responses, diminished performance because of fatigue, and mood and behavioral changes (Mullington et al., 1996). Although astronauts frequently use sleep medication, it has been

BOX 2-2

Altered Sleep Patterns as Example of Multifactorial Problems Arising During Space Travel

Impairment of normal sleep patterns can erode cognitive performance and vigilance during spaceflight (Berry, 1969). Astronauts experience sleep disruption, as shown by a survey of 48 astronauts and polygraphic recordings (Stogatz et al., 1987; Santy et al., 1988; Gundel et al., 1993; Monk et al., 1998), with an average deficit of just under 2 hours per day from the programmed time, with a change in the architecture of sleep, and with less time spent in stages 3 and 4 of sleep. Sleeping pills account for 45 percent of all medications used by space shuttle crews (Putcha et al., 1999).

Sleep problems are multifactorial. They begin with the distorted sleep-wake cycle on launch day and continue with the schedule geared to Mission Elapsed Time rather than local times at takeoff or landing sites. The schedule is further disrupted to accommodate the arrivals and departures of Earth-to-orbit vehicles. In 1990, NASA issued guidelines (Mission Operations Directorate, 1990) limiting the time that the daily sleep schedule can be changed in each day. The 4- to 7-hour phase advance shift is accomplished gradually (20 to 40 minutes earlier per day on space shuttle missions STS-90 and STS-95) or in a series of 2-hour jumps.

Light is the most powerful synchronizer of the human circadian pacemaker (Czeisler and Wright, 1999). There is an external 90-minute light-dark cycle on orbiting flights, which can disrupt the circadian pacemaker and interfere with sleep. Measurements of light levels on Spacelab and Spacehab showed much higher levels on the flight deck than the middeck, with some high readings in the early evening. Since inappropriately timed light exposure or insufficiently intense lighting can be disruptive, astronauts coming from the dimmer middeck to a presleep period on the flight deck may have difficulty with sleep.

The carryover effects of the cyclic use of hypnotics and caffeine can intensify any cognitive deficit and should be avoided. Additional research studies aboard the space shuttles and the ISS will further characterize the nature of sleep disturbances and contribute to the development of safe and effective countermeasures for associated performance decrements.

Abstracted from Barger and Czeisler, 2000.

only modestly effective in correcting the insomnia. Use of melatonin as a possible countermeasure has been discussed, but assessment of effective medication for sleep has not been undertaken in a methodical way. Electroencephalographic changes have been recorded with prolonged wakefulness, and these correlate with neurobehavioral performance capability (Dijk et al., 1992; Dinges et al., 1997).

Clinical Research Opportunity 7. Building a coordinated clinical research program that addresses the issues of neurological safety and care for astronauts during long-duration space travel.

Clinical Research Opportunity 8. Performing pharmacological trials with dose-response and pharmacokinetic measures to assess the efficacies and toxicities of medications commonly used to treat sleep disturbances during space travel.

Clinical Research Opportunity 9. Considering clinical trials on the use of growth hormone and other countermeasures and developing devices to control ambient light and the core temperature at appropriate levels during space travel to reduce sleep disturbances.

Eye-Hand Coordination and Sustained Gross Motor Activity

There is evidence of the degradation of task performance over time during space travel (Manzey and Lorenz, 1998). The tasks studied have included fine motor function, eye-hand coordination (including documentation of the inability to pilot a vehicle for days after landing from short flights), and gross motor activity, especially extravehicular activity. Explanations for this deterioration of performance include difficulty in vigilance, sensory input diminution, and changes in visual acuity (Billica, 2000; Marshburn, 2000a). Neurological recovery of balance and mobility requires 1 to 3 days after short space missions and 10 to 30 days after longer missions.

Few data that can shed light on the causes of these observations are available. Astronauts commonly complain of sluggishness, impaired cognition, and disorientation. In fact, the most frequently identified symptom during space travel is fatigue or asthenia. Some of this may be related to sleep-circadian rhythm disturbance; some of it may have to do with the functioning of the stimuli from the peripheral to the central nervous system. An important issue is sustained performance during long-duration missions in space. Pharmacological countermeasures have addressed sleep but not vigilance or performance-enhancing strategies. Shift assignments have been used to provide more autonomy and more rest and relaxation, with some success. Concerns remain about the possible sequelae of changes in visual acuity, vigilance, balance, and muscular function.

PERIPHERAL NERVOUS SYSTEM

Published reports and presentations to the committee made little mention of sensorimotor status during or upon the return from space missions, yet sensorimotor status is an important component of muscle activation.

The lack of stimulation may be contributing to the decreased muscle mass and possibly the asthenia and fatigue that occur during spaceflight. Devices that stimulate sensory input (vibrators, pneumatic pumps) could be studied to better understand the role that sensorimotor status might play in muscle activity.

Because nervous system health is important to crew health and mission success, data could be collected from studies on the ISS on the efficacies of pharmacological agents in improving nervous system performance. Systematic study of sensorimotor status in microgravity is important to understanding the contributions of neurological and behavioral factors to human performance in deep space.

REPRODUCTIVE SYSTEM

The committee was unable to find many data on the effects of microgravity on the reproductive system. This is understandable, given that space missions thus far have been of relatively short duration. As the lengths of missions are extended, however, and outposts in the solar system occupied by humans become a possibility, the study of the effects of space travel on human reproductive physiology, the risks associated with exposure of the gametes to radiation, and the study of reproduction in space are warranted.

Effects of Radiation on Gametes

Effects on Male Gametes

Spermatogonia are among the most radiosensitive cells in the body. Exposure to radiation at levels as low as 10 REM (roentgen equivalent in man) has been known to cause reduced levels of sperm production (ICRP, 1969), and exposure to levels of 50 REM may cause temporary sterility. A single exposure sufficient to produce permanent azospermia would be fatal to the individual. Exposure to a lower dose of radiation over a protracted period of time, however, would not be lethal but could produce sterility (Jennings and Santy, 1990). Although infertility is an important issue, the long-term genetic consequences of exposure of the gonads to radiation are of greater concern to many.

Effects on Female Gametes

Unlike the testis, which constantly replaces spermatogonia, the ovary has a fixed supply of oocytes that have been present since birth. Because the oocytes do not actively undergo mitotic division and since they possess effective enzymatic repair systems, the ovary is more resistant to radiation-induced genetic effects; however, the effect of radiation on oocytes is cumulative. A single dose of 300 to 400 rads has been sufficient to eliminate oocytes from the ovaries as well as estrogen production. When the exposure is fractionated, there is increased tolerance over a single dose, and primary oocytes show a degree of recovery from the accumulated radiation damage (ICRP, 1969). Assessment of the effects of space travel on the reproductive endocrine system and on ovulatory function should be ongoing, and NASA should consider offering preflight gamete cryopreservation for men and women who may wish to reproduce after a long-duration space mission.

Clinical Research Opportunity 10. Determining whether radiation exposure during space travel causes genetic damage or altered fertility in men and women and, for women, premature ovarian failure.

Human Reproductive Physiology in Space

Male Reproductive Physiology

Although little is known about the effects of space travel on the hypophyseal-pituitary-gonadal axis, some evidence for reversible testicular dysfunction has been found. Reductions in testosterone levels have been reported during flight and postflight in both rats and humans (Plakhuta-Plakutina, 1977; Tigranjan et al., 1982; Deaver et al., 1992). Since testosterone secretion by the Leydig cells in the testis is stimulated by luteinizing hormone, a decrease in testosterone levels may indicate a disturbance in the hypophyseal-pituitary-gonadal axis. If this is the case, one might expect abnormalities of germinal epithelial function and, thus, diminished sperm production as well.

Clinical Research Opportunity 11. Determining female and male reproductive hormone levels during space travel.

Female Reproductive Physiology

At birth, women possess all the gametes that they will ever have. Thus, as women age, so do their gametes, and advanced oocyte age is associated with infertility. Many women, including women who are astronauts, delay childbearing until after they have completed their education and have achieved some of their career objectives. The average age of women selected into an astronaut class is 32 years, and many have not had children. Of 99 female finalist candidates examined during five selection cycles between 1989 and 1997, only 18 had been pregnant. Once they are admitted into the space program, given the constraints of training on a pregnancy, female astronauts commonly further delay childbearing until after the completion of one or two spaceflights. As a result, female astronauts are often in their 40s when they attempt pregnancy (Jennings and Baker, 2000).

Retrograde menstrual flow is considered an etiologic factor in the development of endometriosis and consequent infertility (Sampson, 1927; Scott et al., 1953; Jennings and Baker, 2000). Many women experience some retrograde menstruation that, at Earth's gravity, is usually confined to the pelvis. However, in microgravity there is concern that the level of retrograde menstrual flow might be increased and that instead of being confined to the pelvis it would disperse throughout the abdominal cavity. Abdominal symptoms, shoulder pain, or an obvious reduction in the amount of menstrual flow has not been observed among women during spaceflights. However, retrograde menstrual flow has not been subjected to systematic study (Seddon et al., 1999).

Because of the short durations of space missions so far, coupled with the pulsatile nature of hormone secretion by the hypophyseal-pituitary-ovarian axis, the effect of space travel on ovulatory function has not been studied (Seddon et al., 1999; Strollo, 1999). The effects on the menstrual cycle of stress and exercise during space travel also have not been studied (Seddon et al., 1999). On Earth, stress and exercise can be associated with anovulation, and continuous estrogen exposure can be associated with endometrial hyperplasia and excessive vaginal bleeding. Alternatively, stress and excessive exercise can be associated with hypogonadotropic hypogonadism, resulting in reduced estrogen levels and amenorrhea. The latter condition is associated with decreased bone mineral density. The effect of microgravity, in addition to the effects of stress and an exercise program necessary to maintain cardiovascular and musculoskeletal well-being on long-duration flights, may increase the risk of developing one of these conditions. Exogenous hormone therapy in the form of oral contraceptives or some other therapy may be both preventive and therapeutic.

Clinical Research Opportunity 12. Determining the effect of microgravity on menstrual efflux and retrograde menstruation.

Sex Differences

The National Space Biomedical Research Institute held a workshop on sex-related issues in spaceflight research and health care in August 1999 (Seddon et al., 1999). Focusing on sex-related issues, the group reviewed existing demographics and epidemiological information, identified areas of needed research, and identified ways to accelerate research on the ground and in space to ensure the health of space crews and to provide the best medical care to diverse crewmembers.

The report from the workshop indicates that in most areas there are insufficient data from which to draw valid conclusions about sex-specific differences in physiological responses among astronauts. One of the reasons for the lack of data is the relatively small female astronaut population compared with the size of the male astronaut population, which precludes performance of a study with sufficient statistical power for adequate analyses. In addition, in many areas, the individual differences in physiological responses among members of the same sex are as great or greater than those among individuals of different sexes. This adds to the difficulty in doing the analysis, which occurs in many areas of research involving astronaut health because of the small sample sizes. This is an important research problem in clinical trials with small numbers of participants, such as those involving astronauts (IOM, 2001e).

Other than orthostatic hypotension after space shuttle missions, in which women have a greater likelihood of presyncope during postmission "stand" tests than men (Fritsch-Yelle et al., 1996), the workshop panel members identified few areas in which there were differences by sex. The report points out that there are no data for several systems (postmenopausal bone loss, iron intake, muscle strength, and endurance) in which one might expect sex differences to exist. The report lists other areas in which the prediction of sex differences is not possible and for which no data exist but that are worthy of study. These include susceptibility to decompression sickness, pharmacokinetics and pharmacodynamics, immune function, sensitivity to radiation, and psychosocial adaptation. These are important areas of concern that are also addressed elsewhere in this report.

In addition to identifying areas of concern for female astronaut health care, including pregnancy and the pharmacodynamics of oral contraceptives, the panel looked at issues dealing with the human-machine interface that may affect women during space travel. Apparently, existing equipment, including extravehicular activity suits and shuttle egress suits, and task design preclude some individuals from performing some tasks (Seddon et al., 1999). A stated goal described in the workshop report of Seddon and colleagues (1999) is assurance that all individuals selected to be astronauts are able to perform all tasks associated with the job, regardless of their size or sex. The panel asked that a firm commitment be made to achieving this goal. The present IOM committee supports that recommendation.

Clinical Research Opportunity 13. Collecting clinical data for both men and women when anatomically possible and physiologically sensible for all individuals in the space program and, on a regular basis, subjecting the data to analysis for sex-related differences.

URINARY SYSTEM

Renal stone formation secondary to bone calcium mobilization and excretion in the urine are well-identified concerns of space travel, with an expected incidence of 0 to 5 percent. The effects of microgravity on the urinary system also include changes in urodynamics (unknown incidence) and urinary hesitancy (reported in seven cases). Significant changes in the pH of urine and in urine calcium and citrate levels increase the risk of renal stone formation.

Countermeasures for urinary problems are primarily oriented toward the prevention of nephrolithiasis through adequate hydration. The recommended daily fluid intake volume for astronauts during spaceflight is greater than 2.5 liters (Lane and Schoeller, 2000). Therefore, the availability of a more than adequate supply of water must be ensured so that crewmembers do not hesitate to drink adequate volumes of water to prevent the formation of renal calculi. This may necessitate daily monitoring of water consumption levels, as adequate in-mission treatment of calculi during extended missions may be impossible. As ultrasound devices become smaller, it is possible that an ultrasound or other imaging device will be standard medical equipment for all long-duration space missions. This would make it possible and desirable to perform intramission screening for nephrolithiasis to identify those requiring increased hydration levels to prevent the growth of calculi.

As effective countermeasures for the problem of bone mineral density loss in microgravity are developed, it must be ensured that the solutions to the problem do not result in significantly increased rates of renal stone formation secondary to alterations in calcium metabolism.

PHYSIOLOGICAL MONITORING

NASA has recognized not only the need to measure various parameters to evaluate astronauts during space travel but also the need to store blood and urine for future analysis upon the return to Earth (SSB and NRC, 2000). Both are important requirements for the clinical research program for long-duration space travel. Because of the unique environmental stresses of space travel (gravity forces, radiation), when a new technology is developed, there must be an adequate lead time to test the new technology in the space environment. With the evolution of new technology, NASA will continually be challenged to evaluate the newest methods and instruments in a microgravity environment. For example, the rapid application of nanotechnology, nonivasive biosenors, new imaging techniques, and informatics will provide diagnostic and treatment assessment capabilities vastly different from those available today. Validation of new technologies in the space medicine clinical research program and a transition to the routine application of those technologies in the space medicine clinical research program must be high priorities and continuous challenges for NASA.

Monitoring During Space Travel: Development of Technology

Priority should be given to the development of high-resolution, high-precision, yet minimally invasive or noninvasive methods for the monitoring of important physiological parameters and for biological imaging during all periods of space travel. Technologies are evolving rapidly and will dramatically alter monitoring capabilities during space missions. Technology may be used to minimize the intrusiveness of testing by developing means of obtaining transcutaneous measurements or substituting urine or saliva samples for blood samples. Examples of technologies under commercial development are noninvasive blood glucose monitoring kits and kits for the determination of complete blood counts. The application of nanotechnology to human fluid analysis, genomic analyses, and pharmacogenetics will also likely be routine aspects of future space travel. NASA's hosting of periodic international nanotechnology conferences to bring engineers and biologists together and the agency's collaborative initiative in cell biology with the National Cancer Institute are appropriate means of gaining wider expertise to understand and deal with methods that can be used to monitor risks to human health during future expeditions beyond Earth orbit.

Research Opportunity 14. Giving priority to high-resolution, high-precision, yet minimally invasive or noninvasive methods for the monitoring of physiological parameters and for imaging of the human body during space travel.

A STRATEGY FOR A SPACE MEDICINE CLINICAL RESEARCH PROGRAM

The Critical Path Roadmap project is NASA's integrated, cross-disciplinary strategy to assess, understand, mitigate, and manage the risks associated with long-term exposure to the space environment. Initiated during 1997 and 1998, it is an iterative approach of review, analysis, and deliberations among intramural and extramural investigators focused on a worst-case scenario: long-duration, highly autonomous interplanetary missions such as a human expedition to Mars. The project consists of seven elements (risks, risk factors, critical questions, risk mitigation, the Critical Path Roadmap project, requirements, and deliverables [countermeasures]), as illustrated in Box 2-3. Fifty-five risks have been identified and stratified (Table 2-2). The goal is to identify and validate clinical interventions (countermeasures) for half of the risks by 2006 and for all of the risks by 2010 (Charles, 2000). Continuing revision of the Clinical Path Roadmap is available at *http://criticalpath.jsc.nasa.gov.*

Eleven discipline risk areas have been identified:

- advanced life support;
- environmental health;
- radiation effects;
- human performance;
- bone loss;
- cardiovascular alterations;
- food and nutrition;
- muscle alterations and atrophy;
- immunology, infection, and hematology;
- neurovestibular adaptation; and
- space medicine.

BOX 2-3
Elements of the Critical Path Roadmap Project

✓ **Risks:** Likelihood of an undesirable event occurring as a result of exposure to the space environment. Risk assessment includes the probability of the risk's occurrence, the severity of the consequences of the occurrence, and the current status of mitigation of the risk.

Examples: Occurrence of serious cardiac dysrhythmias
Fracture and impaired fracture healing

✓ **Risk Factors:** A condition or precipitating factor that must be present for the risk to occur. Such conditions can operate singly or in combination to contribute to the occurrence of risk.

Examples: Poor nutrition
In-flight work schedule overload

✓ **Important Questions:** What research and technology need to be developed to further assess the risk and address its mitigation?

Example: Will bone mass loss continue unabated for missions longer than 6 months, or will it plateau at some time consistent with absolute bone mineral density?

✓ **Risk Mitigation:** Strategies, devices, interventions, and requirements that need to be in place to modify the occurrence or impact of the risk.

Examples: Resistive exercise regimens
Medications
Preflight fitness requirements

✓ **Critical Path Roadmap Project:** Graphic representation(s) of the essential set of research and technology development tasks required to address the risks associated with exploration-class space missions with humans, specifically denoting

➡relationships (causal pathway among risks, risk factors, and consequences)
➡priorities
➡temporal sequencing (predecessor questions or tasks)

✓ **Requirements**

Examples: Crew screening and selection requirements for minimization of bone loss
Habitability requirements
Nutritional requirements
Medical care systems for diagnosis, monitoring, and treatment of illness and injury

✓ **Deliverables:** Specific end items that can be identified, completed, and available at known dates.

Examples: Validated, preflight training countermeasures (see Table 2-2) for psychosocial adaptation
Validated, in-flight virtual reality-based training for emergency egress
Exercise countermeasures for bone loss, muscle atrophy, and cardiovascular and neurovestibular adaptation

Source: Charles, 2000.

TABLE 2-2 Critical Path Roadmap Project: Critical Risks. SOURCE: Charles, 2000

Bone Loss	Cardiovascular Alterations	Human Behavior and Performance	Immunology, Infection and Hematology	Muscle Alterations and Atrophy
Acceleration of age-related osteoporosis		Human performance failure because of poor psychosocial adaptation		
Fractures (traumatic, stress, avulsion), and impaired healing of fractures	Occurrence of serious cardiac dysrhythmias	Human performance failure because of sleep and circadian rhythm problems		Loss of skeletal muscle mass, strength, or endurance
	Impaired cardio-vascular response to orthostatic stress			Inability to adequately perform tasks due to motor performance problems, poor muscle endurance, and disruptions in structural and functional properties of soft and hard connective tissues of the axial skeleton
				Inability to sustain muscle performance levels to meet demands of performing activities of various intensities

Neurovestibular Adaptation	Radiation Effects	Clinical Capability	Other	
	Carcinogenesis caused by radiation	Trauma and acute medical problems		**Severe Risks**
Disorientation and inability to perform landing, egress, or other physical tasks, especially during/after g-level changes	Damage to central nervous system from radiation exposure	Toxic exposure	Inadequate nutrition (malnutrition) (three risks)	**Very Serious Risks**
Impaired neuromuscular coordination and/or strength	Synergistic effects from exposure to radiation, microgravity and other environmental factors	Altered pharmacodynamics and adverse drug reactions	Postlanding alterations in various systems resulting in severe performance decrements and injuries	
	Early or acute effects from radiation exposure		Habitation and life support (eight risks)	

continued

TABLE 2-2 Continued

Bone Loss	Cardiovascular Alterations	Human Behavior and Performance	Immunology, Infection and Hematology	Muscle Alterations and Atrophy
Injury to connective tissue or joint cartilage, or intervertebral disc rupture with or without neurological complications	Diminished cardiac function	Human performance failure because of human system interface problems and ineffective habitat and equipment design, etc.	Immuno-deficiency/infections	Propensity to develop muscle injury, connective tissue dysfunction, and bone fractures due to deficiencies in motor skill, muscle strength, and muscular fatigue
Renal stone formation	Manifestation of previously asymptomatic cardiovascular disease	Human performance failure because of neurobehavioral dysfunction	Carcinogenesis caused by immune system changes	
	Impaired cardiovascular response to exercise stress		Altered hemo- and cardio-dynamics from altered blood components	Impact of deficits in skeletal muscle structure and function on other systems
			Altered wound healing	
			Altered host-microbial interactions	
			Allergies and hypersensitivity reactions	

Neurovestibular Adaptation	Radiation Effects	Clinical Capability	Other	
Impaired cognitive and/or physical performance due to motion sickness symptoms or treatments, especially during/after g-level changes	Radiation effects on fertility, sterility and heredity	Illness and ambulatory health problems		**Serious Risks**
Vestibular contribution to cardioregulatory dysfunction		Development and treatment of decompression illness complicated by microgravity-induced deconditioning		
Possible chronic impairment of orientation or balance function due to microgravity or radiation		Difficulty of rehabilitation following landing (two risks)		

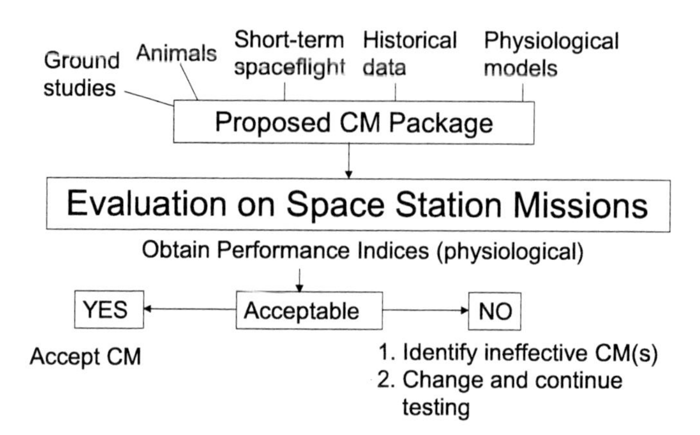

FIGURE 2-2 Countermeasure (CM) development and evolution. Sources: Feiveson, 2000; IOM, 2001e.

These risk areas have been grouped into four categories:

- environmental and technological,
- human behavior and performance,
- human health and physiology, and
- medical care capabilities.

It is expected that clinical research will be an integral component in the successful prevention or mitigation of the risks to astronaut health in each of the discipline risk areas. The Critical Path Roadmap project, with its clinical research program on countermeasures, is NASA's mechanism for the development of priorities and validation of the means of mitigating the risks that humans may be expected to encounter during exploration-class missions into deep space. As a mechanism for gathering information, the Critical Path Roadmap project is an appropriate model for clinical research in space medicine, with use of the ISS as the critical platform for clinical research in space medicine and for NASA to validate countermeasures in the space travel environment as it prepares to go beyond Earth orbit (Figure 2-2). NASA has, however, not integrated a prospective and continually reevaluated understanding of clinical research on astronaut health into the Critical Path Roadmap project in any of the information that it provided to the committee or that the committee was able to find. It should be noted that the environment on Mars, at 0.4G and with some atmosphere, is different from that of the spacecraft. Whether the presence of significant gravitational pull will tend to ameliorate the adverse effects of microgravity is unknown but is worth consideration in the planning of research.

A Clinical Research Program for NASA

IOM has conducted a number of studies focused on various aspects of clinical research (IOM, 1998, 1999a, 2001c,d,e). Although there are many definitions of clinical research (IOM, 1994), IOM recently endorsed the broad definition of clinical research set forth by the Graelyn Conference and National Clinical Research Summit (Box 2-4) and formed the Clinical Research Roundtable to periodically discuss continuing developments and needs in clinical research.

The varied, significant, and potentially harmful changes in physiology associated with space travel require health care interventions to protect the well-beings of the participating astronauts and the integrity of the mission. However, traditional clinical research with astronauts is difficult because of the small number of participants available for study, the inability to replicate microgravity and its effects on Earth, restrictions on the use of control groups, and limitations on the substitution of results from studies with animals.

A large portion of NASA's human clinical research to date has been in the form of countermeasure development and evaluation, and this research has been conducted through trials of single interventions directed at specific physiological changes. This largely precludes, however, the classic approach of comparing alternative treatments in clinical trials with adequate statistical power (adequate sample sizes for statistical analysis) to demonstrate safety and efficacy (IOM, 2001e). Individual variations in both physiological

BOX 2-4
What Constitutes Clinical Research?

Clinical research embraces a continuum of studies involving interaction with patients, diagnostic clinical materials or data, or populations in any of these categories: disease mechanisms; translational research; clinical knowledge, detection, diagnosis, and natural history of disease; therapeutic interventions, including clinical trials; prevention and health promotion; behavioral research; health services research, including outcomes; epidemiology; and community-based and managed care-based research.

Source: AAMC, 1999, p. 4.

changes and responses make it even more difficult to evaluate results because of the small number of participants (i.e., astronauts). Because of these difficulties, NASA has had to select the most promising clinical intervention for its studies of countermeasures to avoid further dilution of its study group and possible ethical breaches.

In addition to this empirical approach, more basic research is sorely needed to define the mechanisms that produce the physiological changes that result from space travel, which, it is hoped, will improve the focus of countermeasure clinical trials. The present set of NASA priorities may account for the limited success of countermeasures to date. This does not imply deficiencies on the part of NASA researchers but recognizes the severe limitations under which they work.

Potential methods for improvement in clinical research with small numbers of participants are described in a recent IOM study report, sponsored by NASA, *Small Clinical Trials: Issues and Challenges*, which reviews study designs and statistical methods for determination of the validities and efficacies of studies with small numbers of participants (IOM, 2001e). NASA's request for the study is one example of the agency's attempt to reach out in new directions to better use the limited opportunities during space missions to conduct clinical research with astronauts.

Formation of the National Space Biomedical Research Institute and increased collaboration with the National Institutes of Health (e.g., the new initiative with the National Cancer Institute) are additional directions that will provide NASA with wider expertise.

It is not clear, however, how well clinical research design, strategies, and so forth fit into these initiatives, as clinical research with astronauts relevant to space medicine beyond Earth orbit, particularly validation of countermeasures, can largely be conducted only during space missions. There does not appear to be a clearly visible and transparent plan of who oversees, designs, reviews, and is responsible for this clinical research. NASA's organizational charts, lines of authority and responsibility, administration, and budgeting processes and its levels of accountability presented to the committee (in official NASA policy documents and during briefings by NASA officials throughout the committee's information-gathering process) describe a fragmented, nonuniform process of strategic planning and oversight that is insufficient to provide an effective means of understanding and mitigating the risks to human health from traveling in space.

A comprehensive strategic plan for the prevention and amelioration of the many potential risks to well-being during long-duration travel beyond

Earth orbit and treatment of the conditions that result from those risks is vital to the future of space medicine. This requires an understanding of the relationship between human physiology and adaptation to space travel. It also requires a broad, evidence-based, collaborative, and coordinated clinical research effort to ensure that NASA has the ability to provide health care commensurate with the expectations of the medical community and society at the time of launch beyond Earth orbit. Some of the clinical research opportunities in space medicine are listed in Box 2-5.

CONCLUSION AND RECOMMENDATION

Conclusion

NASA has devoted insufficient resources to developing and assessing the fundamental clinical information necessary for the safety of humans on long-duration missions beyond Earth orbit.

- *Although humans have flown in space for nearly four decades, a paucity of useful clinical data have been collected and analyzed. The reasons for this include inadequate funding; competing mission priorities; and insufficient attention to research, analysis including insufficient investigator access to data and biological samples, and the scientific method.*
- *Although NASA's current approach to addressing health issues through the use of engineering design and countermeasures has been successful for short-duration missions, deep space is a unique environment that requires a different approach.*
- *A major problem of space medicine research is the small number of astronaut research participants, which requires special design and analysis of the data from clinical trials with small numbers of participants. This necessitates a strategy focused on maximizing opportunities for learning.*

Recommendation

NASA should develop a strategic health care research plan designed to increase the knowledge base about the risks to humans and their physiological and psychological adaptations to long-duration space travel; the pathophysiology of changes associated with environmental forces and disease processes in space; prediction, development, and validation of preventive, diagnostic, therapeutic, and rehabilitative

BOX 2-5
Clinical Research Opportunities for Astronaut Health

Musculoskeletal System

1. Establishing the course of changes in bone mineral density and markers of bone mineral density turnover in serum and urine before, during, and after space travel.
2. Developing a capacity for real-time measurement of bone mineral density and enhanced three-dimensional technology to assess the risk of fracture during space travel.
3. Identifying human phenotypes and genotypes resistant to space travel-induced bone mineral density loss.
4. Tailoring therapeutic interventions (i.e., countermeasures such as diet, exercise, and medications) as a high priority and then validating the promising countermeasures in studies with astronauts during exposure to microgravity.

Cardiovascular

5. Considering artificial gravity and pharmacological interventions as solutions to orthostatic hypotension.

Gastrointestinal

6. Investigating the relationship between space motion sickness and absent bowel sounds including pharmacological and adaptive countermeasures.

Nervous System

7. Building a coordinated clinical research program that addresses the issues of neurological safety and care for astronauts during long-duration space travel.
8. Performing pharmacological trials with dose-response and pharmacokinetic measures to assess the efficacies and toxicities of medications commonly used to treat sleep disturbances during space travel.
9. Considering clinical trials on the use of growth hormone or other countermeasures and developing devices to control spacecraft ambient light and the core temperature at appropriate levels during space travel to reduce sleep disturbances.

Reproductive Health

10. Determining whether radiation exposure during space travel causes genetic damage or altered fertility in men and women and, for women, premature ovarian failure.
11. Determining female and male reproductive hormone levels during space travel.
12. Determining the effect of microgravity on menstrual efflux and retrograde menstruation.

Physiological Monitoring

13. Collecting clinical data for both men and women when anatomically possible and physiologically sensible for all individuals in the space program and, on a regular basis, subjecting the data to analysis for sex-related differences.
14. Giving priority to high-resolution, high-precision, yet minimally invasive or noninvasive methods for the monitoring of physiological parameters and for imaging of the human body during space travel.

measures for pathophysiological changes including those that are associated with aging; and the care of astronauts during space missions.

The strategic research plan should be systematic, prospective, comprehensive, periodically reviewed and revised, and transparent to the astronauts, the research community, and the public. It should focus on

- **providing an understanding of basic pathophysiological mechanisms by a systems approach;**
- **using the International Space Station as the primary test bed for fundamental and human-based biological and behavioral research;**
- **using more extensively analog environments that already exist and that have yet to be developed;**
- **using the research strengths of the federal government, universities, and industry, including pharmaceutical, bioengineering, medical device, and biotechnology firms; and**
- **developing the health care system for astronauts as a research database.**

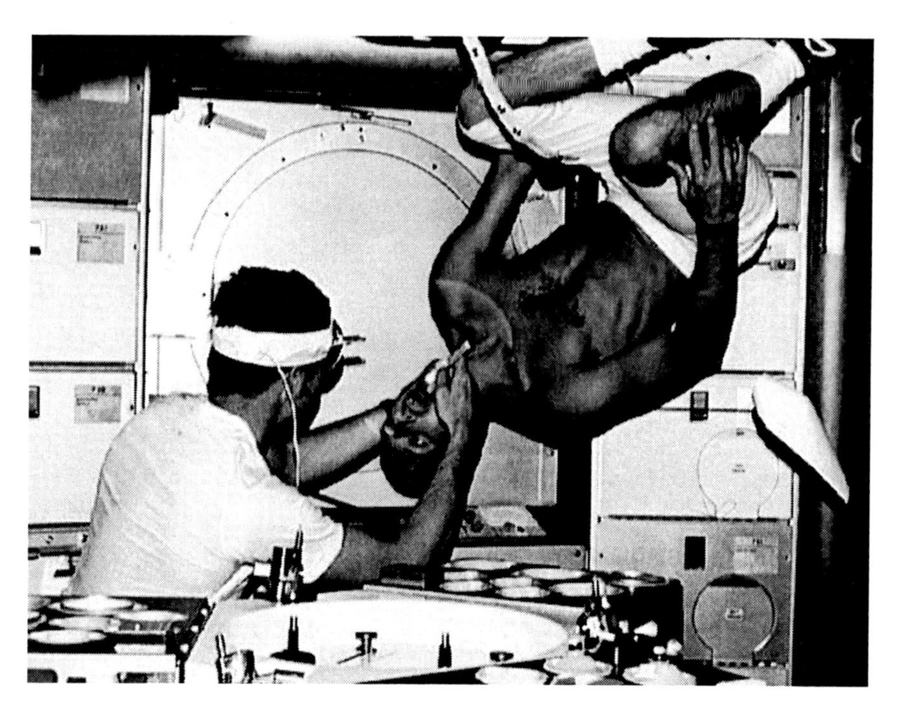

Astronaut Charles Conrad, Jr., commander of the first manned *Skylab* mission, undergoing a dental examination by Medical Officer Joseph Kerwin, M.D., in the *Skylab 2* Medical Facility during Earth orbit on June 22, 1973. In the absence of an examination chair, Conrad simply rotated his body to an upside down position to facilitate the procedure. NASA image.

3
Managing Risks to Astronaut Health

We had to struggle a little bit, but we showed the reason that manned spaceflight has been as successful as it has for a large number of years. A large team of people scattered across the entire planet were able all together to get a major advance in the space station assembly operations.

Chuck Shaw, lead flight mission director for the 100th shuttle mission commenting from Houston Mission Control on the successful attachment by the shuttle *Discovery* astronauts of the nine-ton Z1 structural truss to the International Space Station's Unity module, October 14, 2000

Perhaps the most ambitious goal of the National Aeronautics and Space Administration's (NASA's) space medicine program is to be able to provide optimal health care to the first (and subsequent) astronauts who will go on exploration-class missions to Mars. Because any such mission lies more than a decade in the future, the challenge to the Institute of Medicine (IOM) Committee on Creating a Vision for Space Medicine During Travel Beyond Earth Orbit was this: what can usefully be said, so far in advance, about providing day-to-day health care in space, while on the Martian surface, and during the return trip to Earth? What types of illnesses and injuries might reasonably be anticipated on long-duration space missions?

In this chapter, the committee tries to begin answering those questions by looking at the only evidence available: the morbidity and mortality experiences of U.S. astronauts and Russian cosmonauts, U.S. Navy submariners, and Australian scientists and explorers in the Antarctic. This look back includes findings from physical examinations conducted in space to see what is normal, or baseline, in microgravity. The committee also examines potential health problems in each of several medical practice areas—cardiology, neurology, surgery, and psychiatry, to name a few—in which critical risks have been identified (see Table 2-2).

The committee anticipates that long-duration missions beyond Earth orbit will be qualitatively different from short spaceflights. Medical and behavioral issues that have not been particularly problematic on short flights may loom large on exploration-class missions. It is not possible to accurately predict the treatment innovations, technological advances, and shifting standards of care that may occur over the next 20 years and prove relevant to medical practice in space.

GENERAL PRINCIPLES AND ISSUES

The focus of this chapter is the care of the individual patient in space. Premission evaluation should include assessments of both the astronaut's health status (including the status of specific organ systems at risk, such as the musculoskeletal system [see Chapter 2 for the risks involved]) and other risks. The general principles of care are the maintenance of normal health status in microgravity and, if illness or injury occurs, restoration of normal function as quickly and efficiently as possible during and upon the return from the space mission. As part of a responsible space crew, each crewmember should be expected to participate in routine surveillance to be able to measure the health status of other members of the crew at regular intervals. *Resources should be available for the diagnosis and treatment of the most common minor and major illnesses and injuries that are anticipated to occur in the Earth environment, as well as to diagnose and treat conditions that are unique to microgravity and the particular space mission. The crew should be prepared to treat a wide variety of conditions of various degrees of severity during a space mission and, most of all, be prepared to treat the unexpected.*

The major health and medical issues related to exploration-class missions have been of little risk or concern to NASA up to the present for short-duration space travel (e.g., space transportation system [STS] space shuttle missions) (Box 3-1). All of the major health and medical issues are projected, however, to be moderate to severe concerns that affect astronaut health on the International Space Station (ISS), and except for radiation protection and bone mineral density loss, the degree of severity of the other health and safety challenges have yet to be estimated for exploration-class missions. Many of these issues and challenges are directly related to or are completely tied to known human physiological adaptations to space travel. Separation of these issues from the discussions of physiological adaptations in Chapter 2 is in many cases artificial. Similar concerns, issues, and topics on medical, surgical, rehabilitative, and behavioral health in this chapter and in Chapters 4 and 5 must also be considered in the continuum of clinical research and health care for astronauts to begin building the infrastructure and health care system (Chapter 7) needed for human exploration of deep space. The committee has chosen to separate these topics into chapters to place the emphasis on clinical research (Chapters 2 to 5), health care (Chapters 3 to 5), and opportunities and ethical and infrastructure concerns (Chapters 6 and 7) that it believes is necessary to promote the needed attention to the safe passage and the health of astronauts during travel beyond Earth orbit and into deep space.

BOX 3-1
Major Health and Medical Issues During Spaceflight

Health or Medical Issue	GRD	AIR	STS	ISS	EXP
Radiation protection	G	G	G	Y	R
Hearing conservation	G	G	G	R	TBD
Cardiovascular	G	G	G	Y	TBD
Muscle	G	G	G	Y	TBD
Bone loss	G	G	G	Y	TBD
Neurovestibular	Y	NA	G	R	TBD
Habitability	NA	G	Y	Y	TBD
Extravehicular activity risk	NA	G	Y	Y	TBD
Medical care	Y	NA	Y	Y	TBD
Diversity (age, gender, etc.)	Y	NA	Y	Y	TBD
Psychological issues	Y	G	G	Y	TBD
Workers' compensation	Y	G	G	Y	TBD

Abbreviations: GRD, ground; AIR, airflight; STS, space shuttle; ISS, International Space Station; EXP, exploration-class mission; G, green, little or no risk; Y, yellow, moderate risk; R, red, severe risk; TBD, to be determined; NA, not applicable.

Source: Williams, 2000.

Medical Events in Extreme Environments

Evidence Base from Previous Space Missions

A review of 79 U.S. space missions involving 219 person-flights lasting 2 to 17 days each (Putcha et al., 1999) reported that the most common conditions experienced were space motion sickness (SMS), nasal congestion, and sleep disorders. None of these medical conditions have required the mission to end, have been life threatening, or have required intensive medical treatment; they are bothersome but are not medical emergencies. Exploration-class missions, however, because of their lengths of as many as 3 years beyond Earth orbit, raise in NASA's current judgment the probability of a major medical event, a condition requiring intervention by a medical practitioner, during the mission (Billica, 2000).

A study of 175 astronauts from 1959 to 1991 reported 20 deaths (19 males and 1 female), mostly unrelated to spaceflight because of high rates of automobile and aircraft accidents and accidental deaths on the *Apollo 1* and the *Challenger* spacecrafts. The small numbers of participants and the premature deaths from injuries may well mask the morbidity and mortality figures from other disorders related to spaceflight, such as cancer, if the participants live long enough (Peterson et al., 1993). Related disorders such as the development of cancer and cardiovascular, arthritic, and other conditions may increase in frequency as the duration of space travel and the ages of astronauts increase, just as they would had the same individuals remained on Earth.

The risks of medical events increase with the lengths of missions (Billica et al., 1996). A survey of the perception of risk from spaceflight was returned by 65 medical professionals and showed that medical events with the highest perceived likelihood of occurrence had the least effect on the mission or the crew, but those with the greatest impact on the mission or crew were least likely to occur (Billica et al., 1996). Skin disorders (irritation from fiberglass, contact dermatitis, rashes, and furuncles) were thought to be the most common, followed by respiratory and digestive disorders.

NASA reported that 1,867 medical events occurred from 1981 to 1998 on space shuttle flights STS-1 to STS-89 (Billica, 2000). Among the population of 508 individuals on those flights, 498 reported a medical event or symptom other than SMS. The events, derived from a histogram presented to the committee (Billica, 2000), were ill-defined symptoms ($n = 788$), respiratory events ($n = 83$), symptoms related to nervous system or sensory organs ($n = 318$), digestive disorders ($n = 163$), symptoms related to skin or subcutaneous tissue ($n = 151$), symptoms related to the musculoskeletal system ($n = 132$), and injuries ($n = 141$). Approximately 5 percent (77 of 1,777) were injuries, and 10 deaths occurred, 7 during a catastrophic explosion in the early phase of the launch (*Challenger* in 1986) and 3 from a fire on the launchpad (*Apollo 1* in 1967).

Rates of events have not been reported, and associations of illness or injury with extravehicular activity (EVA) also have not been reported. EVAs are associated with a high workload and are associated with a much higher risk of injury because of the momentum imparted to large masses during EVAs and the lengthy periods of work outside the spacecraft (Nicogossian et al., 1994). Even in the non-EVA microgravity environment, fractures are possible due to movement of cargo, which can easily "get away" once set in motion or if an individual pushes away from a wall too hard and experiences a bone-jarring hit on the opposite wall (Nicogossian et al., 1994). This is a good example of the importance of training to prevent medical injury.

The microgravity environments of long-duration space missions will also be associated with overexertion, strains, and sprains, because backaches and effects from the physical demands of EVAs have been reported during shorter missions and require pharmacological treatment (Putcha et al., 1999). Backaches are not specifically associated with EVAs but are a common complaint thought to be associated with elongation in vertebral column length and stress placed on intervertebral discs. This type of pain has been reported to be ameliorated by axial compression performed by crewmembers while in orbit (NASA, 2000b).

The medications administered as a single dose or taken by only one person during 219 space missions have included phenazopyridine, omeprazole, zolpidem, sucralfate, an antifungal (Vagisil), clotrimazole (Mycelex), docusate, an antacid (Gaviscon), cimeditine, diclofenac, meclizine, ofloxacin, gentamicin, lovastatin, flavoxate, ketoprofen, metaxalone, and cephalexin. This spectrum of medications that has been taken and the disorders that have been treated on short missions point to the need to plan for a broad-based and space medicine-focused pharmacopoeia to treat a wide variety of signs, symptoms, and diseases on longer missions. The existence of such a pharmacopoeia also necessitates procedures to avoid potential abuse.

It would have been helpful to the committee's assessment if the data on illness and injury with and without an association with EVA made available to the committee had been stratified and publicly reported to allow the committee and others to have a better understanding of the health-related risks of spaceflight. Moreover, a number of questions remain unanswered. In addition, facts needed to best appreciate any list of health-related risks of spaceflight (Tables 3-1, 3-2, 3-3, and 3-4) to plan for future space missions were not available. For instance, (1) how were the symptoms distributed among astronauts with different specialties (e.g., pilots versus payload specialists)? (2) did astronauts who flew more than one mission experience fewer symptoms on subsequent flights? and (3) what was the degree of severity of the reported symptoms?

It is therefore important to look at the totality of the data from space missions and what has been learned from other extreme isolated environments on Earth (e.g., Antarctica and extended underwater submarine missions). These data relate to the type and incidence of medical-surgical and behavioral health events that occur in these environments and are needed to best gauge and plan for future needs during extended space travel before commencement of exploration-class space missions with astronaut crews.

Evidence Base from Extended-Duration Submarine Missions

Medical events during submarine missions are instructive as they occur in a confined, remote environment where there is limited diagnostic and therapeutic support. They occur in an atmosphere where potentially life-threatening or other severe medical illnesses can end a mission, in the sense that the submarine is required to interrupt or even abort its mission.

The U.S. Navy described the incidence of illnesses and injuries on 136 submarine patrols from January 1, 1997, through December 31, 1998. The numbers of acute encounters were related to the total number of person-days under way, with 2,044 acute encounters in 1.3 million person-days at sea, or 157 acute encounters per 100,000 person-days (Table 3-5). Stratified by illness and injury, illness accounted for 112.9 episodes per 100,000 person-days, with 70 percent able to maintain full duty; and accidents accounted for 37.2 episodes/100,000 person-days, with 55 percent able to maintain full duty (Thomas et al., 2000).

A different perspective is obtained when the health disorders and medical-surgical procedures in Table 3-5 are compared with the reasons for medical evacuations from U.S. submarines (Table 3-6). A range of 1.9 to 2.3 medical evacuations per 1,000 person-months was reported for all submarines in the U.S. Atlantic Fleet from 1993 to 1996. A range of 1.8 to 2.6 evacuations per 1,000 person-months was reported for humane reasons (i.e., death or serious illness in the family) (Sack, 1998), suggesting that if these

TABLE 3-1 In-Flight Medical Events for U.S. Astronauts During the Space Shuttle Program (STS-1 through STS-89, April 1981 to January 1998)

Medical Event or System by ICD-9[a] Category	Number	Percent	Incidence/14 days
Space adaptation syndrome	788	42.2	2.48
Nervous system and sense organs	318	17.0	1.00
Digestive system	163	8.7	0.52
Skin and subcutaneous tissue	151	8.1	0.48
Injuries or trauma	141	7.6	0.44
Musculoskeletal system and connective tissue	132	7.1	0.42
Respiratory system	83	4.4	0.26
Behavioral signs and symptoms	34	1.8	0.11
Infectious diseases	26	1.4	0.08
Genitourinary system	23	1.2	0.07
Circulatory system	6	0.3	0.02
Endocrine, nutritional, metabolic, and immunity disorders	2	0.1	0.01

[a]International Classification of Diseases, 9th edition.

SOURCE: Billica, 2000.

data are extrapolated to extended space travel or habitation, the psychosocial support needs may well be just as important as the medical needs in a long-duration space mission.

The medical reasons for submarine evacuations from 1993 to 1996 varied (Table 3-6). The largest number of conditions requiring medical evacuation are trauma and "other" (miscellaneous). It should be noted, however, that psychiatric reasons rank second in the specific categories. The "other" category most likely consists of large numbers of unrelated clinical conditions, further reinforcing the diversity of clinical conditions that can be expected to occur during a space mission. Factors such as astronaut age and medical prescreening would affect the incidence of medical emergencies among the members of the space crew, but since prescreening for most conditions cannot be done, it is possible that similar disorders and the proportions of those disorders that could occur among the members of a space crew would be similar to those that occur among individuals on submarine missions.

TABLE 3-2 Medical Events Among Seven NASA Astronauts on *Mir*, March 14, 1995, through June 12, 1998

Event	Number of Events	Incidence/100 Days
Musculoskeletal	7	0.74
Skin	6	0.63
Nasal congestion, irritation	4	0.42
Bruise	2	0.21
Eyes	2	0.21
Gastrointestinal	2	0.21
Psychiatric	2	0.21
Hemorrhoids	1	0.11
Headaches	1	0.11
Sleep disorders	1	0.11

NOTE: Data from the Russian Space Agency reports that there were 304 in-flight medical events onboard the *Mir* from February 7, 1987, through February 28, 1998. The numbers of astronauts at risk or the incidence per 100 days was not reported.

SOURCE: Marshburn, 2000b.

Tansey and colleagues (1979) reviewed health data from 885 Polaris submarine patrols from 1963 to 1973, for 4,410,000 person-days of submarine activity. They described 1,685 medical events that resulted in 6,460 duty days lost. Only events that resulted in the loss of at least 1 workday were reported. The events with the six highest rates of occurrence were, in descending order, trauma, gastrointestinal disease, respiratory infections, dermal disorders, infection, and genitourinary disorders. The spectrum of disorders was very broad and included cases of arrhythmia, paroxysmal superventricular tachycardia, infectious hepatitis, gastrointestinal hemorrhage, meningococcemia, paranoid schizophrenia, appendicitis, pilonidal abscess, perirectal abscess, ureteral calculi, testicular torsion, and crush injuries, further emphasizing that the scope of anticipated medical conditions on long-duration space missions will be very broad (Tansey et al., 1979).

The incidence of the types of illnesses observed during extended submarine missions is generally similar to the incidence encountered during spaceflights. NASA has used the incidence of medical events on submarines to estimate that there may be one major medical event requiring intervention of the type usually delivered by a medical practitioner during a future exploration-class mission of 3 years in length with five to seven astronauts (Billica, 2000; Flynn and Holland, 2000). Unfortunately, the nature of that

TABLE 3-3 Medical Events and Recurrences Among Astronauts of All Nationalities on *Mir,* March 14, 1995, through June 12, 1998

Event	Number of Events	Recurrences
Superficial injury	43	2
Arrhythmia	32	98[a]
Musculoskeletal	29	NR[b]
Headache	17	8
Sleeplessness	13	9
Fatigue	17	4
Contact dermatitis	5	3
Surface burn	5	NR
Conjunctivitis	4	2
Acute respiratory infection	3	NR
Asthenia	3	2
Ocular foreign body	3	NR
Globe contusion	2	NR
Dental	2	NR
Constipation	1	NR

[a]See Chapter 2.
[b]NR, not reported.

SOURCE*:* Marshburn, 2000b.

NOTE: Further information on symptom duration, functional impact, or recurrences, especially the nature of arrhythmias and the number of astronauts who experienced them, is important for assessment of the potential impacts of such events on prolonged space missions. Other than arrhythmias, the medical events reported were minor, although most were certainly as vexing in space as they would be on Earth.

event is unpredictable, so preparations must be prioritized and must still be made for a wide spectrum of problems.

Evidence Base from Antarctic Expeditions

The Australian National Antarctic Research Expeditions (ANARE) Health Register compiled 1,967 person-years of data from 1988 to 1997. It documents 5,103 illnesses and 3,910 injuries (Table 3-7). The distribution and variety are similar overall to those from spaceflight data.

Seventeen Australians, moreover, have died in the Antarctic and subantarctic since 1947 (Taylor and Gormly, 1997; D. J. Lugg, ANARE, personal communication, August 24, 2000). Excluding those conditions peculiar to

TABLE 3-4 Pharmacopoeia Usage During *Mir* Missions

Medications	Number of tablets or doses dispensed
Pseudoephedrine	131
Zolpidem	81
Temazepam	68
Diphenhydramine	60
Aspirin	55
Acetaminophen	37
Bisacodyl	32
Ibuprofen	28
Terfenadine	18
Long-acting phenylpropanolamine	13
Nose drops (Neosynephrine)	9

SOURCE: Marshburn, 2000b.

NOTE: This list reaffirms the discomforts experienced by crew of previous missions and suggests the probability that nasal congestion, sleep disorders, pain, and constipation will afflict the crews of longer-duration space missions.

TABLE 3-5 Incidence of Health Disorders and Medical-Surgical Procedures During 136 Submarine Patrols

Disorder	Number/100,000 Person-Days
Injury (includes accidents)	48.8
Respiratory	24.6
Skin or soft tissue	19.0
Ill-defined symptoms	10.5
Infections	10.0
Procedure	**Percentage of All Procedures Performed**
Wound care, splinting	42.0
Suturing	18.7
Cleansing	8.2
Nail removal	6.8
Fluorescein eye examination	4.2
Incision and drainage of abscess	2.9
Tooth restoration	2.0

SOURCE: Thomas et al., 2000.

TABLE 3-6 Reasons for 332 Medical Evacuations from All Submarines, U.S. Atlantic Fleet, 1993 to 1996

Reason for Evacuation	Number of Cases
Trauma	71
Psychiatric illness	41
Chest pain	34
Infection	40
Kidney stone	23
Appendicitis	21
Dental problem	31
Other	71
Total[a]	332

[a] Rate = 1.9 to 2.3 per 1,000 person-months.

SOURCE: Sack, 1998.

the Antarctic environment (drowning and exposure, $n = 5$; outdoor injuries, $n = 5$), seven nonpredictable deaths occurred in the Antarctic because of appendicitis ($n = 1$), cerebral hemorrhage ($n = 1$), acute myocardial infarction ($n = 2$), carbon monoxide poisoning ($n = 1$), perforated gastric ulcer ($n = 1$), and burns ($n = 1$). Each of these is a possible medical event on a spacecraft, indicating the wide variety of medical emergencies that can occur and that must be considered in planning for health care management in

TABLE 3-7 ANARE Health Register Illnesses in Antarctica from 1988 to 1997

Disorder	Number	Percent
Injury and poisoning	3,910	42.0
Respiratory	910	9.7
Skin, subcutaneous	899	9.6
Nervous system or sensory organs	702	7.5
Digestive	691	7.4
Infection or parasitic	682	7.3
Musculoskeletal or connective tissue	667	7.1
Ill-defined symptoms	335	3.6
Mental	217	2.3

SOURCE: Lugg, 2000.

BOX 3-2
Potential Methods of Risk Assessment and Screening

Physiological profiling. Physiological profiling consists of profiling of the central nervous system; cardiovascular system; pulmonary system; musculoskeletal system; eyes, ears, nose, and throat; gastrointestinal system; and genitourinary system and gynecological health and profiling for endurance, strength, and adaptive ability.

Psychosocial profiling. Psychosocial profiling evaluates individuals for moods or traits that are positively selected and traits that lead to rejection. Are there behavioral signs and symptoms that validly and reliably predict maladaptive or disruptive behavioral interactions?

Health status profiling. Past medical events and current signs and symptoms are used as part of health status profiling. Individuals with disorders such as migraine and hypertension and individuals who have had keratoplasty are excluded from astronaut training.

Markers to assist in profiling. Such markers include the results of laboratory studies, genetic profiling, and imaging studies and family history.

the extreme environment of extended-duration space travel or habitation beyond Earth.

Health Risk Assessment

Individuals differ in their susceptibilities to disease, vulnerabilities to environmental assaults, and abilities to recover from injury. Current potential methods of risk assessment and screening (Box 3-2) rely heavily on the identification of preclinical disease or conditions known to predispose an individual to illness. For example, certain abnormal lipid profiles are an identified risk factor for atherosclerosis, elevated blood pressure is an identified risk factor for stroke and heart disease, and osteoporosis is a risk factor for hip fracture.

As a result of the Human Genome Project, investigators are also identifying DNA sequences that correlate with an increased risk for a particular disease or syndrome, and the basis for this increased risk is being elucidated. The presence of DNA or RNA sequences indicative of a potential health risk as well as those indicative of a preferential pharmacodynamic or other response to treatment may be an integral part of the standard of care in the future. Breast cancer (Box 3-3) and colon cancer are two diseases for which there are well-established risk assessment and screening tools, as well as an increasing number of DNA sequences that indicate a propensity for an increased risk of development of the diseases. As such linkages become more prevalent for a wider variety of diseases, they will offer increased means for profiling and screening of individuals to decrease major medical and health risks (Box 3-1) and promote the health and safety of astronauts for long-duration space travel.

While on a long-duration mission beyond Earth orbit the starting point for medical care will most often be a description of the chief complaint and a physical examination. The physical examination technique must be adapted to the microgravity environment, where the method of determination of internal organ location and other diagnostic methods differ widely from those used in terrestrial environments. The patient, examiner, and equipment must be stabilized for proper examination technique in microgravity. The examiner must learn adaptive movements to perform the abdominal examination properly. Auscultation of heart or bowel sounds in the noisy spacecraft environment is difficult, and stethoscopes need to be modified with this in mind by the cooperative work of engineers and clinicians. Harris and colleagues (1997) have reported a number of variations to physical examinations performed in a microgravity environment (Box 3-4).

The physical findings listed in Box 3-4 are for five males and two females during 8- and 10-day missions (Harris et al., 1997). Facial and periorbital edema are most evident during the first 3 days of flight but persist throughout the mission. Facial and periorbital edema, nasal congestion, and jugular venous distention occur because of fluid shifts to the head and torso because the loss of gravity eliminates hydrostatic pooling in the extremities. Thinning of the lower extremities is due to an approximately 40 percent reduction in interstitial fluid levels in the lower extremities (Baisch, 1993).

Auscultated bowel sounds diminish on the 2nd through 5th days of the flight. Investigators note that absent bowel sounds were strongly associated with the development of motion sickness. The diaphragm is elevated by two intercostal spaces and appears to remain elevated during spaceflight. Normal variations in the positions of thoracic and abdominal viscera also occur in microgravity. It is important to recognize these variations because they could affect the interpretation of physical findings and the performance of invasive or diagnostic procedures. For example, failure to recognize a "normal" elevation of a hemidiaphragm in microgravity could result in improper placement of a thoracostomy tube for the treatment of pneumothorax, or alterations of peritoneal signs in microgravity could change the signs and

BOX 3-3
Breast Cancer as an Example of Risk Assessment in Space Medicine

The possibility that breast cancer can occur in a woman while on long-duration space travel exists. The development of breast cancer in a physician at a base in Antarctica in 1999 reinforces the potential that a breast mass including cancer could develop while an astronaut was in space. Breast cancer will have been diagnosed in 182,000 women in the United States in the year 2000 (American Cancer Society, 2000). The risk of breast cancer increases with advanced age, early age at menarche (younger than 12 years), nulliparity or late age of first birth (older than 30 years), a family history of breast cancer in a mother or sister, and breast biopsies showing proliferative disease or atypical hyperplasia (Harris, 1991; Harris et al., 1992). Women who have had thoracic irradiation or who carry an inherited DNA mutation predisposing them to breast cancer are at even greater risk. Up to 10 percent of all cases of breast cancer are believed to be associated with an inherited predisposition for the disease (Claus et al., 1996). It is estimated that 55 to 85 percent of women carrying the *BRCA-1* mutation will develop breast cancer during their lifetimes (Ford et al., 1998). In addition, carriers of the mutation tend to develop breast cancer at a younger age. Women with strong family histories can be tested for mutations in the *BRCA-1* and *BRCA-2* genes. If they carry a mutation, the risk for breast cancer is increased severalfold. More than 50 percent of carriers are diagnosed with breast cancer by age 50 (Easton et al., 1995).

Individuals who are homozygous for mutations in the ataxia telangiectasia gene (*AT*) are at significantly increased risk of cancer and are highly sensitive to ionizing radiation. It is not currently known if exposure to high-energy radiation such as that which occurs beyond Earth orbit will also increase the risk. An increased risk of breast cancer has been noted in *AT* heterozygotes (Athma et al., 1996), which constitute approximately 1 percent of the general population in the United States and Europe, and it has been hypothesized that early and frequent mammography may be inadvis-

symptoms of appendicitis. Pelvic examination has not been reported in microgravity, but a pelvic examination in microgravity may lead to other variations in normal findings.

Health Care Opportunity 1. Expanding, validating, and standardizing a modified physical examination, the microgravity examination technique, and including a technique for pelvic examination for use in microgravity.

Nutrition

Food quality and variety affect crew attitudes and overall performance. Nutritional concerns include sufficient caloric intake, nutritional density, food palatability, varied menus, and cultural variations in preferred foods. It able for this group of individuals. Because breast cancer is so common in the general population, there is tremendous ongoing research that may be relevant to NASA and that must be monitored to obtain an understanding of the risk due to exposure to ionizing radiation as well as to high-energy particle radiation.

In 1989, Gail and colleagues published the Gail model for estimation of an individual's risk for the development of breast cancer over time on the basis of age, age at menarche and age at first birth, family history, and breast biopsy results. By using this model, after entry of the risk factor for a given woman, both her 5-year and lifetime probabilities for breast cancer can be calculated (Gail et al., 1989; Benichou et al., 1997). Women who have had thoracic irradiation or a specific family profile indicative of an inherited DNA mutation, however, cannot be appropriately assessed with the Gail model.

The average age of women selected for the space program is 32 (Jennings and Baker, 2000). Given the astronaut training constraints on pregnancy, however, once female astronauts are admitted into the program, they commonly defer childbearing. Therefore, because many female astronauts delay childbirth, they are at increased risk for breast cancer. The increased risk imposed by radiation while in space is unknown, but it may be an additional risk. The probability that breast cancer would develop to the point of clinical significance in a woman while on a long-duration space mission therefore exists. To what extent should preflight risk assessment include laboratory testing of an individual's nucleic acid composition as well as conventional screening? Should an individual identified to be at increased risk be eliminated from a long-duration space mission beyond Earth orbit or counseled and allowed to make an informed decision? These are questions for NASA to address in its preparation for exploration-class missions.

A similar scenario can be presented for colon cancer in men and women, particularly as the average age of astronauts increases, and will become more common for other disorders as the understanding of the molecular biology of disease continues its present rapid advance.

is critical that the food supply be adequate, safe, and reliable and that it remains so throughout the mission. Inadequate food and water supplies or contamination or loss of the supplies, particularly since much will have to be generated from recycled materials during the mission, will result in termination of the mission or the loss of life.

Additionally, one must consider methods that can be used to ensure the adequacy of caloric intakes to prevent the ongoing loss of body mass. On the basis of current experience (Lane and Schoeller, 2000), a degree of malnutrition is anticipated in nearly all astronauts during space travel without the use of countermeasures and is expected in even 61 to 94 percent of astronauts with the use of countermeasures. Just as the effects of zero gravity or microgravity on the pharmacodynamics and metabolism of pharmaceuticals are unknown (see below), absorption of nutrients may be problematic, lead-

BOX 3-4
"Normal" Findings on Physical Examination in Microgravity

Facial and periorbital edema
Oily facial skin
Hyperemia: facial skin, conjunctivae, mucosae of the nose, and mucosae of the pharynx
Jugular venous distention
Elevation of diaphragms by two intercostal spaces
Point of maximal cardiac impulse displaced substernally or not palpable
Posture: barrel chest, hyperextended back, flexion of upper and lower extremities
Extremities: thinning of lower extremities
Neurological: hyperreflexia

Source: Harris et al., 1997.

ing to unexpected deficiencies that result in the need for supplementation. Nutritional requirements have been found to be similar for short-duration space missions and life in normal terrestrial environments, but energy intake is decreased during space travel, so most astronauts lose body mass, including 1 to 2 liters of body water. A monitored mandatory caloric intake may be considered, as may monitoring of nutritional status in more standard ways, for example, via measurement of arm circumference. One must consider that inadequate intakes of micronutrients or vitamins would adversely affect the entire crew, making identification of all required nutrients and their absorption or elimination pharmacodynamics a priority.

Pharmacodynamics and Pharmacokinetics

Pharmacodynamics deals with the interactions of drugs and living systems, whereas pharmacokinetics is the study of the absorption, distribution, and metabolism-utilization of pharmacologicals. The microgravity environment can be expected to affect the pharmacodynamics and pharmacokinetics of all drugs, yet little clinical research has been performed in these areas. Clinical research on the pharmacokinetics and pharmacodynamics of drugs in space is limited by the small numbers of participants, limited opportunities for clinical study (i.e., few space missions), and the lack of a reasonable terrestrial proxy for microgravity in which to conduct pharmacological studies.

Drugs administered in microgravity may not have the anticipated local, regional, or systemic effects and may manifest different adverse effect profiles in space compared with those observed on Earth. For example, a case series of 21 crewmembers given 25 to 50 mg of promethazine intramuscularly reported only a 5 percent sedation rate, whereas the sedation rate was 60 to 73 percent in studies conducted in standard Earth gravity (Bagian and Ward, 1994). This phenomenon needs to be closely studied for several reasons. The decreased effectiveness of a sedative could be due to SMS or the sheer excitement associated with the space mission. There is also some evidence that receptor interactions may be altered under conditions of hypovolemia (Derendorf, 1994). The bioavailabilities of oral drugs given in space can be affected by gastric emptying, gastric motility, and hepatic blood flow (Tietze and Putcha, 1994). Bed rest, which is sometimes used to partially simulate the effects of microgravity, is reported to delay the absorption of common oral medications, and drug distribution is affected by the redistribution of fluids from the lower body to the head and torso in space (Tietze and Putcha, 1994). The bioavailabilities of oral scopolamine and acetaminophen are altered in flight and may be affected by SMS and the particular day of the mission (Cintron et al., 1987; Tietze and Putcha, 1994). Drug binding by protein and tissue is presumably altered in microgravity because of muscle and tissue atrophy, the latter of which has been documented upon the return from a space mission (Edgerton et al., 1995).

The frequency of use of medications during spaceflight (Putcha et al., 1999) is such that targeted research into the pharmacokinetics of various routes of drug administration (oral, intranasal, transcutaneous, subcutaneous, intramuscular, intravenous) is required, with the goal of determining the predictability of the effect and efficacy. The resources for the medical crew on the spacecraft for a long-duration mission should include a compendium of the indications and adverse effects of the pharmaceuticals on board and their anticipated kinetic changes, such as bioavailabilities and half-lives, that are predicted for the microgravity environment.

Health Care Opportunity 2. Developing an easily accessible database for medications on the spacecraft, including dosage, indications, adverse effects, and anticipated changes in the pharmacokinetic profile in microgravity.

Environmental and Occupational Health

Environmental Hazards

The environmental and occupational health of astronauts will be important issues for long-duration space travel. Missions beyond Earth orbit will dictate a unique set of requirements to protect crewmembers from hazards such as chemical contamination, volatile organic compounds, particulate matter, and microorganisms. Crewmembers may confront the challenge of living in a noisy environment, where vibration is also a potential hazard to human health and to sensitive experiments (Koros, 1991a; Koros et al., 1993). Crewmembers will work in an environment of artificial light, which could adversely affect their performance (Czeisler et al., 1990; Barger and Czeisler, 2000). Missions may include scheduled and unscheduled EVAs, which will be physically challenging and conducted by humans who may be physiologically compromised. Finally, there is the question of the deleterious effects of exposure to primary and secondary radiation (Johnston and Dietlein, 1977; Nicogossian and Parker, 1982; SSB and NRC, 2000). Operational requirements for EVAs will present significant physical challenges for crews. For the ISS, an estimated total of 1,100 hours will be required to carry out planned construction and maintenance. At a high inclination of orbit of 51.6 degrees, such activity will expose the crew to high-altitude radiation as well as temperature extremes, micrometeors, and physical injuries.

Decompression Sickness

Decompression sickness (DCS) represents another significant potential threat to astronauts on long-duration missions. Should emergency EVAs or sudden unexpected decompression of the spacecraft occur, DCS might ensue. NASA is well aware of these problems and is actively pursuing solutions to these issues, particularly so that it can effectively and safely finish the construction of the ISS. The committee believes that NASA's efforts in these areas should continue by including investigations of the possible relationship between a patent foramen ovale and DCS. Careful integration of engineering issues and habitability should take place in planning for EVAs as well as emergency contingencies for EVAs. Finally, the committee is confident that advances in materials will allow the inclusion of a lightweight, low-volume recompression chamber in the manifest for missions beyond Earth orbit, if it is considered necessary (i.e., if a pressure suit with an internal pressure greater than that in present suits has not been developed).

Internal Environment

During long-duration space missions, the internal environment of the spacecraft will offer its own unique challenges, ranging from chemical and microbial contamination to noise and vibration. Like any confined habitat there will be chemical and physical toxic elements. On a spacecraft, however, the crew has few opportunities to replace or recondition a toxic environment.

Exposure to Toxic Chemicals Environmental hazards come from several sources. Propulsion propellant (Freon, hydrazines, nitrogen dioxide) leaks into the spacecraft interior can be toxic in small quantities (Tansey et al., 1979). The spacecraft crew can be further exposed if propulsion chemicals enter the spacecraft through the air lock or if they crystallize on EVA suits. Accidental chemical releases during space shuttle flights have also been reported. The dominant source appears to be heat degradation of electronic devices. Thermodegradation of spacecraft polymers with the production of formaldehyde and ammonia adds to the environmental hazard (Nicogossian et al., 1994). There were nine incidents from STS-35 to STS-55, with four resulting from burning electrical wiring. Subsequent analysis found benzene, acetaldehyde, dimethyl sulfides, and other compounds in the space shuttle crew compartment atmosphere (James et al., 1994; Pierson et al., 1999).

The experience on *Mir* provided a glimpse of the potential risks of contamination from the very systems designed to protect the health of the crew. The Freon in cooling loops presented a significant hazard to the crew when the Freon was released. Oxygen canisters in *Mir* presented a life-threatening problem for the crew when they caught fire (Burrough, 1998; Linenger, 2000). There was danger not only from fire but also from the smoke and particulate matter released from the canisters themselves.

Data collected during the Extended-Duration Orbiter Medical Project's evaluation of volatile organic compounds in the cabin atmosphere indicated that levels were below maximum allowable concentration (SMAC) limits in the spacecraft. It was noted that most pollutants reach a state of equilibrium within the first 3 to 4 days of a mission; however, the exceptions are hydrogen, methane, dichloromethane, and formaldehyde. Dichloromethane and formaldehyde are of concern because both have significant toxic properties.

Missions of 2 weeks' duration measured dichloromethane levels of 0.79 milligrams per cubic meter (mg/m^3; 30-day SMAC of 20 mg/m^3) and formaldehyde levels as high as 0.08 mg/m^3 (30-day SMAC of 0.05 mg/m^3) (Pierson et al., 1999).

Exposure to hazardous materials during space travel could result in multiple casualties with serious injuries, burns, or smoke inhalation that would soon outstrip the finite resources available on the spacecraft. Planning to minimize exposure of the crew includes the identification of potential hazards, recognition that a hazardous material is responsible for acute signs and symptoms, identification of the agent(s) involved, retrieval and review of information regarding toxicity and secondary contamination, protection of unexposed personnel from primary and secondary contamination, methods for triage and decontamination of the exposed individual(s), and treatment of the injured and exposed individuals. Available resources should be modified as the technology advances and should be easily available to the crew (ATSDR, 1991; Sidell et al., 1991). Methods for continuous surveillance for toxic contaminants should be in place, using Earth analog models (ATSDR, 1997).

Health Care Opportunity 3. Developing an easily accessible hazardous materials manual for space travel to aid in the surveillance, detection, decontamination, and treatment of chemical exposures.

The concentrations of the particulate pollutants detected in the space shuttle ranged from 35 to 56 mg/m^3, with the majority of them being greater than 100 micrometers in diameter. Most particles did not settle out of the atmosphere during the mission. Most were organic in nature and were most likely generated by crewmembers (Pierson et al., 1999).

Health Care Opportunity 4. Monitoring and quantifying particulates on a continuing basis.

Microbial Contamination The quantification of airborne bacteria and fungi indicates that the levels of bacteria increase moderately with the duration of the mission and that the levels of fungi decrease with the duration of the mission. The levels of bacteria range from a few hundred to 1,000 colony-forming units per cubic meter (CFU/m^3) of air during longer missions. Fifteen species of bacteria were recovered from samples collected during space missions. *Staphylococcus*, *Micrococcus*, *Enterobacter*, and *Bacillus* species were found on 85 percent of the missions; and *Staphylococcus aureus* was recovered during 57 percent of the missions (Nicogossian and Parker, 1982).

Fungi tended to be present at a few hundred CFU/m^3 early in the missions, but their quantities dropped to undetectable levels toward the ends of the missions. Nevertheless, low levels of *Aspergillus* and *Penicillium* species are found during greater than 60 percent of the missions (Nicogossian and Parker, 1982; Mehta et al., 1996).

The essential questions are as follows: How transmissible are these organisms? How much mixing of flora occurs between and among crewmembers? Is bacterial or fungal overgrowth an issue for long-duration space travel? Lastly, how does the radiation environment affect the growth of microorganisms and their toxicities to humans?

Health Care Opportunity 5. Examining the capability of microbial identification, control, and treatment during space travel.

Noise In both the U.S. and the Russian space programs, noise has been a major problem. Spacecraft noise levels disrupt sleep, increase stress and tension, and can result in temporary or even permanent hearing loss. The environmental control system, system avionics, and payload experiments generate most of the noise. The design limits of most work environments range from 63 to 68 decibels (dBA). The noise levels on a number of space missions has exceeded this baseline limit, exposing the crew to noise levels far greater than normal terrestrial noise levels. The maximum permissible continuous exposure level in a work environment is 90 dBA for an 8-hour period, according to the Occupational Safety and Health Administration. Early in the life of the ISS it was about 75 dBA. In a spacecraft, the environmental noise level is steady, and it continues for months. This will certainly have a deleterious affect on astronauts' hearing (Koros, 1991b; Koros et al., 1993), and it may affect astronauts' concentration and behavior as well.

Health Care Opportunity 6. Developing methods for noise cancellation or reduction.

Ergonomic Issues Because the human body has evolved on Earth in the presence of Earth's gravity astronauts are vulnerable to ergonomic problems in microgravity. During a space mission, crewmembers try to maintain a neutral body posture, wherein the shoulders, arms, hips, and legs are flexed and in a relaxed position. Working on an experiment such as one in a glove box, however, requires the crewmember to work against the natural tendency of the body to assume this posture. This increases fatigue, decreases performance, and predisposes crewmembers to injury (Mount and Foley, 1999).

Health Care Opportunity 7. Standardizing ergonomic practices on the basis of the human body's response to the microgravity environment.

External Environment

Radiation Much is known about the radiation environment of low Earth orbit; little, however, is known about the radiation environment of high Earth orbit and beyond. The ISS will be exposed to primary ionizing radiation and high-energy particle radiation from solar and galactic sources (SSB and NRC, 2000a). Exposure to secondary radiation—that is, radiation emitted from spacecraft metals and other materials following collision of their nuclei with high-energy solar or galactic particles penetrating the spacecraft shell—may also be a problem. In low Earth orbit, a band of atmospheric radiation, known as the Van Allen belts, is concentrated over the South Atlantic, hence the term South Atlantic Anomaly. An orbiting spacecraft will spend only 2 to 5 percent of its time in this region, but astronauts receive more than half of their total radiation doses during this period.

Most penetrating radiation from the Sun results from solar particle events (SPEs) and mostly consists of protons generated by solar storms. The Earth's geomagnetic field shields against solar particle events up to 6,370 kilometers above the Earth (Letaw et al., 1987, 1988). Recent reports note that construction of the ISS will take place during a period of maximum solar activity, when the probability of encountering SPEs and Earth-trapped radiation is high. During periods of intense solar activity, solar winds result in elevated intensities of energetic electrons. These are known as known as highly radioactive events. Galactic cosmic rays make up about a third of the radiation in space and produce a continuous low-level form of radiation. Protons of only 10 million electron volts (MeV) of energy can penetrate a space suit, and 25- to 30-MeV protons can penetrate the space shuttle (Lemaire et al., 1996). Beyond Earth orbit, the issue of radiation exposure presents a major challenge for NASA as the quantities of solar and galactic radiation and the potential for exposure increase.

Health Care Opportunity 8. Developing methods to measure human solar and cosmic radiation exposures and the means to prevent or mitigate their effects.

HEALTH CARE PRACTICE OPPORTUNITIES

Cardiovascular Care

Cardiovascular integrity is essential to the health and well-being of astronauts on long-duration space missions, but there is no experience with the delivery of cardiovascular care on such missions. Therefore, the information and recommendations presented here are derived from the few published data on cardiovascular complications incurred during the Mercury, Gemini, Apollo, and Skylab missions (SSB and NRC, 1998c; Charles et al., 1999) and from general principles of cardiovascular care on Earth (Braunwald, 1999).

Standards for the initial screening of astronauts, follow-up annual physical requirements, and causes of rejection are listed in NASA's *Astronaut Medical Evaluation Requirements Document* (NASA, 1998a). Although there are provisions for waivers, it is reasonable to assume that selected crewmembers are at low risk for the development of cardiac problems. This is important, because it will be extremely difficult to treat moderate to severe cardiovascular complications during a long-duration space mission. Space limitations will preclude an extensive pharmacy or medical procedure unit, and no individual with expertise in cardiovascular system-related procedures may be on board to handle cardiovascular complications.

Some of the cardiovascular symptoms and abnormalities that astronauts may present with during a long-duration space mission include high blood pressure with or without symptoms; atrial and ventricular premature beats; atrial arrhythmias, such as atrial fibrillation, atrial flutter, and supraventricular tachycardia; sustained and nonsustained ventricular fibrillation; chest pain, ischemic and nonischemic; shortness of breath, cardiac and noncardiac; orthostatic hypotension; syncope; vasovagal and other cardioneurogenic responses; and edema, cardiac and noncardiac. A strategy must be in place to deal with these on a risk-assessed priority basis and with the possible occurrence of myocardial infarction, whose incidence may increase with the generally increasing age of astronauts at the start of space missions and the extended lengths of missions beyond Earth orbit.

Physiological adaptation to planetary gravity after long-term exposure to microgravity may take several days to weeks, with considerable individual variability. Symptoms from adaptive conditions such as orthostatic hypotension, whether on the Moon or Mars or after the return to Earth, should be treated as needed, with the understanding that normal physiological regulatory mechanisms will take over, allowing physiological function to return to normal.

Health Care Opportunity 9. Providing a thorough cardiovascular evaluation similar to the premission evaluation at the cessation of space travel to provide useful data as part of the continuum of astronaut care and to aid in establishing an evidence base for cardiovascular disorders during space travel.

Dental Care

In 1978, Soviet cosmonaut Yuriy Romanenko experienced a toothache during a 96-day flight of *Salyut 6*. As his problem worsened, Romanenko gulped painkillers and crewmembers pleaded for help from the ground. The Soviet space program had no contingency plans for dental emergencies; the advice from controllers was "take a mouthwash and keep warm." Romanenko, "his eyes literally rolling with pain" (Wheatcroft, 1989, p. 7), suffered for 2 weeks before *Salyut 6* touched down on schedule. His ordeal was the subject of a televised interview in the Soviet Union, as well as published accounts in Russian and U.S. space and dental literature (Wheatcroft, 1989). It also focused attention, including that of NASA, on the need to address the possibility of debilitating dental emergencies in space.

In April 2000, the IOM Committee on Creating a Vision for Space Medicine During Travel Beyond Earth Orbit held a public workshop entitled Space Dentistry: Maintaining Astronauts' Oral Health on Long Missions (see Appendix A). Presentations from invited experts, as well as other data and information reviewed by the committee, suggest that dental problems need not be a major health care issue for astronauts on long-duration missions. This optimistic outlook assumes appropriate premission dental screening and excellent preventive care, as well as the ability to provide in-flight prophylaxis and restorative treatment as needed. A review of advances in preventive dentistry (Box 3-5) led one workshop presenter to predict that by 2020 NASA may be able to select the first Mars crew from a pool of caries-free astronauts (Mandel, 2000).

Still, good teeth and a history of preventive care cannot guarantee that no caries will develop in anyone over the course of a 3-year mission. Some factors that could contribute to the development of tooth and gum disease include changes in bacterial flora in the mouth, inattention to good dental hygiene, changes in food consistency because of the consumption of dehydrated space meals, and lack of foods with natural gingival cleansing proper-

BOX 3-5
Advances in Preventive Dentistry

A three-part preventive strategy, aggressively pursued by dental researchers and practitioners since 1971, may mean a caries-free pool of astronaut candidates for the first mission to Mars with humans. The strategy consisted of the following components:

I. Combating caries-inducing microorganisms
- Recognition of dental caries as an infectious disease.
- Identification of mutans group streptococci as major cariogenic organisms.
- Development of an antibacterial agent, chlorhexidine (also effective against plaque, gingivitis).
- Use of chlorhexidine in a prescription rinse, professionally applied varnish, and self-applied gel.

II. Modifying diet
- A campaign to convince public to restrict sweets was found to be insufficient.
- A nonacidogenic sweetener in chewing gum, xylitol, reduces levels of mutans group streptococci.
- Protective food components (e.g., polyphenols in chocolate) show promise.
- Preservatives with enhanced antibacterial activity are under investigation.
- Natural demineralization inhibitors are under investigation.

III. Increasing resistance of teeth to decay as a result of
- Community water fluoridation.
- Professionally applied fluoride solutions, sealants, gels, and varnishes.
- Self-applied daily fluoride rinses, brush-on gels, and fluoridated toothpastes.
- Controlled-release systems to deliver predetermined amounts of fluoride intraorally.
- Fluoride-releasing dental restorative materials to provide site-specific protection.
- Clinical testing of other remineralizing agents (e.g., amorphous calcium phosphate).
- Testing of new coatings and new coating technologies (e.g., polymeric coatings).
- Study of laser light's ability to alter enamel surfaces, increasing resistance to acid challenge.
- Study of salivary proteins with direct antibacterial properties.

Source: Mandel, 1996.

ties. For these reasons, the crew should be prepared to use restorative techniques and materials in microgravity, and NASA should support the development of new restorative techniques and materials that can be used in microgravity.

The committee's workshop on oral health included a presentation on atraumatic restorative treatment (ART), which may represent one potentially useful approach to the management of dental lesions in space. It is a conservative approach to caries management, in which carious tooth tissue is removed with hand instruments instead of electric rotating handpieces. The cavity is restored (filled) with an adhesive restorative material such as glass-ionomer. The result is a sealed restoration (Estupiñán-Day, 2000).

The reported advantages of ART include little or no pain, reduced need for local anesthesia, minimal trauma to the tooth, conservation of healthy tissue, and simplified infection control. Moreover, ART can be performed by individuals who are not dentists. The technique was originally devised for use in developing countries and disadvantaged communities where access to high-quality, definitive dental care is problematic. The Pan American Health Organization is evaluating the longevity of glass-ionomer restorations under various conditions (Estupiñán-Day, 2000). How they might perform in microgravity is not known, however.

The committee has reviewed dental health data from long-term missions in analog environments. Such data may be of limited predictive value, however. ANARE reported on the dental health experiences of 64 men over a 42-month period. There were 73 reported dental events, which accounted for 8.80 percent of all medical events (Fletcher, 1983). All the men had been prescreened and found to be "dentally fit." The preexpedition screening examinations lacked uniform criteria, however, precluding useful comparisons of that population with other populations. Moreover, the examining dentists did not have the advantage of today's tools for early detection of developing caries.

The committee has learned that NASA is developing new, prevention-oriented dental protocols for space shuttle missions and the ISS and that these are undergoing internal agency review (M. Hodapp, NASA, personal communication, April 10, 2000).

An important question remains unanswered: does exposure to microgravity result in the loss of bone mineral density in alveolar bone? To date, no human data bearing on this question have been reported. The question arises because of the well-documented loss of bone mineral density in the weight-bearing bones. Also, although no human data exist in the current database, microgravity-induced decreases in bone density might also contribute to tooth and gum disease.

Health Care Opportunity 10. Developing a program for instruction in basic dental prophylaxis, the treatment of common dental emergencies such as gingivitis, tooth fracture, dental trauma, caries, and dental abscesses; and tooth extractions.

Endocrine Function

Changes in endocrine function in microgravity have been reported, but the clinical significance and effects on adaptation or maladaptation need more research to determine if they have clinical importance during long-duration space missions. The number of subjects is small and the data are sometimes conflicting. The polar tri-iodothyronine (T3) syndrome has been described in persons living for extended periods in Antarctica. It is characterized by baseline elevations of thyroid-stimulating hormone (TSH) levels, exaggeration of increases in TSH levels in response to a challenge with thyroid-releasing hormone, and more rapid production and clearance of T3 and thyroxin but normal levels of both in serum (Reed et al., 1990).

Thyroid axis kinetics should be further studied during space missions (Lovejoy et al., 1999), since models of prolonged bed rest used as analogs for conditions in space have also demonstrated changes in T3 levels and the effects of T3 on nitrogen balance and catabolism. Little has been reported on adrenal function, but its association with sleep disturbances should be investigated, as circadian fluctuations in steroid levels have been well described on Earth (Birketvedt et al., 1999).

Testosterone levels fell in both humans and rats during space missions and on their return to Earth, and studies with rats did not show changes in spermatogenesis (Plakhuta-Plakutina, 1977). There have been no reports that astronauts have had difficulty with reproduction, but no effects from long-duration space mission have been studied (Tigranjan et al., 1982; Deaver et al., 1992).

The metabolic stress syndrome is of great importance to space medicine, and cortisol production via adrenocorticotropic hormone production by the pituitary gland has been used as a marker. Cortisol levels increase during the first 2 days of a space mission, as do the rates of protein turnover and acute-phase protein synthesis (Stein et al., 1996), documenting the stress of launch and entry into orbit. Caution must be exercised in interpreting data on endocrine function for humans in space because of the large variations in hormone levels among humans, the problems of collection and storage of samples, and the variabilities of assays.

Gastrointestinal Issues

Gastrointestinal problems account for 8 percent of the recorded medical events on space shuttle missions (Billica, 2000). The incidence is 0.52 per person per 14 days in the space environment. Experience in analog environments suggests that the incidence of gastrointestinal problems is much lower, being only 0.01 per person per year. These data suggest that the motility problems identified during space shuttle missions can be attributed to the effects of the microgravity environment. Many astronauts who develop symptoms of SMS also seem to develop a transient ileus, diagnosed by an absence of bowel sounds. Although motility may remain decreased throughout the space mission and the bacterial population may change, the etiologies are unclear and data from short-duration space missions do not suggest that these lead to significant medical problems.

Some spacecraft crewmembers have experienced constipation during missions. This may be related to physiological alterations in the bowel induced by the microgravity environment, but the etiology remains unclear. Adequate hydration throughout long-duration space missions should prevent constipation. Some crewmembers have experienced diarrhea during the later parts of missions. The etiology is also unknown, but it may simply be related to overmedication for constipation. Diarrhea in the space environment presents several problems, including constant use of the Waste Containment System and dehydration, which may exacerbate landing orthostasis. Over-the-counter medications (Imodium and Pepto Bismol) for oral ingestion are available in the Shuttle Orbiter Medical Systems (SOMS) kit. Vigorous hydration with oral or intravenous fluids is recommended. Episodes of diarrhea during long-duration space missions can be treated similarly.

Gastrointestinal problems are largely prevented through optimal premission screening, and most residual symptomatic problems can be treated in a manner similar to that used on the ground. However, three problems to which the gastrointestinal tract is particularly prone are infection, malignancy, and inflammation.

Obstruction of the gallbladder or appendix with calculi and subsequent infection can be lethal without operative intervention. Consideration must be given to prophylactic cholecystectomy and prophylactic appendectomy before long-duration space missions. Although the procedures can be performed today with a minimum of morbidity and a low likelihood of any late postoperative complications, it is not clear whether prophylactic removal is warranted. A careful risk-benefit calculation should be performed with future data. Population-based data to determine how rapidly gallstones can form in an ultrasound-negative patient may be one useful methodology for determination of the advisability of prophylactic cholecystectomy.

Malignancy of the gastrointestinal tract can be ruled out through endoscopy. On the basis of current practice it would appear prudent that astronauts (especially those over age 50) being sent on long-duration space missions have a recent colonoscopy (with or without a concomitant air-contrast barium enema). Consideration might also be given to the screening of candidates by esophagogastroduodenoscopy before long-duration space travel for esophageal, stomach, and duodenal problems.

Inflammation of the pancreas, pancreatitis, is a life-threatening disease even with the best medical care. The many etiologies of pancreatitis include gallstones and specific medications. The use of pharmacological agents that continue to be developed and that are recognized to be associated with the development of pancreatitis should be avoided.

Given the technology available today and in the foreseeable future, it is unlikely that surgical procedures on the gastrointestinal tract (except for percutaneous drainage of an abscess, etc.) will be performed on long-duration missions beyond Earth orbit. At this time development of specific countermeasures related to the gastrointestinal tract does not appear to be required before long-duration space travel.

Gynecological Health Issues

Although the likelihood may be small that a gynecological condition that requires surgical intervention will occur during space travel, the occurrence of such a condition would present special problems. Therefore, as with other medical conditions, attention must be directed toward prevention, conversion of surgical conditions to medically treatable conditions, and, where necessary, the ability to do surgery. For example, pregnancy and its inherent complications, including spontaneous abortion, ectopic pregnancy, and abnormal bleeding, can be prevented with appropriate contraception. Sexually transmitted diseases and their complications will be able to be prevented or treated. Oral contraceptives can help the abnormal bleeding associated with anovulation and the development of functional cysts of the ovary. Surgical conditions, such as uterine myoma, endometriosis, and dysfunctional uterine bleeding, can now be treated medically.

Regardless of these measures, there will be some conditions, such as ovarian neoplasm, adnexal torsion, or bleeding, which will require surgery.

Although elective laparoscopic appendectomy before prolonged space missions has been given some consideration, surgical prevention of adnexal abnormalities is not a consideration except perhaps for postmenopausal astronauts, and the relative risk and treatment means and priority must continue to be evaluated.

Contraception and Hormone Replacement Therapy

Because of the absolute preclusion of pregnancy while in the space program, many female astronauts have chosen contraceptive methods that are known to be very effective. These include intrauterine devices, implants, and oral contraceptives made up of various combinations of hormones, all of which have been continued during space travel (Jennings and Baker, 2000). Although an intrauterine device is an effective contraceptive, for long-duration space missions, the added noncontraceptive benefits of hormonal contraceptive agents may make them preferable. For example, hormonal agents reduce the volume of menstrual flow. Although most women on oral contraceptives have a withdrawal cycle every 28 days, it is possible to extend the cycle from every 28 days to several months. With some hormonal agents, complete cessation of bleeding for the duration of the mission may be possible. Oral contraceptives frequently relieve the dysmenorrhea associated with menses. The ovaries of women on oral contraceptives are less likely to form cysts, which may undergo torsion or other complications. Not only are oral contraceptives an effective way to manage dysfunctional uterine bleeding or bleeding associated with anovulation, should that occur in space, oral contraceptives are also effective treatment for the estrogen deprivation associated with hypogonadotropic hypogonadism and prevention of the bone mineral density loss associated with this condition.

Hormone replacement therapy is known to be effective in preventing osteoporosis among women of certain ages on Earth. During space travel, hormone replacement therapy, as on Earth, will be of importance in preventing calcium loss in postmenopausal crewmembers.

Health Care Opportunity 11. Studying the bioavailability and pharmacological function of exogenous hormone therapy during space travel and, as new medical therapies for gynecological surgical conditions evolve, testing of these therapies for use during space travel.

Hematology, Immunology, and Microbiology

Decreased red blood cell mass during space missions has been recognized since 1977 (Johnson et al., 1977; Leach and Johnson, 1984), but there is no resulting impairment from this "anemia." The documented fall in erythropoietin levels and the fall in the numbers of reticulocytes indicate that this results from diminished production, not increased cellular destruction (Alfrey et al., 1996). However, because of the diminution of the plasma volume early in the flight, the measured hematocrit levels and red blood cell counts did not fall during flight but were noted after the return to Earth because of the more rapid restoration of plasma volume than the level of red blood cell production. Erythropoietin levels returned to normal in 1 to 2 weeks after landing. In 1990, Koury and Bondurant (1990) hypothesized that erythropoietin prevents programmed cell death in erythropoid progenitor cells, thereby adding significantly to general medical knowledge through research conducted in space. Anemia could become a clinical problem during long-duration space travel, and erythropoietin administration is being evaluated as a countermeasure.

Altered cell-mediated immunity has been reported in a variety of analog environments, including the Antarctic (Tingate et al., 1997) and space (Kimzey et al., 1975, 1977). *Escherichia coli* and *Staphylococcus aureus* isolates have also been shown to become more resistant to selected antibiotics during space travel (Lapchine et al., 1986). Although the clinical significance of these alterations has not been determined, the effects on skin and wound infections and wound healing during long-duration space missions could become clinically important.

Preflight isolation techniques for spaceflight crewmembers are reported to have decreased the infection rate for the 3 weeks preflight from about 50 percent to only occasional events (Ferguson, 1977; Ferguson et al., 1977). Gingivitis and skin furuncles are now the primary preflight infections reported (Taylor and Gormly, 1997). Increased shedding of herpesvirus and expansion of Epstein-Barr virus-infected B cells have been reported in the Antarctic environment (Tingate et al., 1997; Lugg and Shepanek, 1999) and in astronauts (Payne et al., 1999). The bases for the divergent changes are not understood. They do, however, indicate the importance of immunology and microbiology to healthy human physiology during space travel and the need for further research in this area of space medicine.

Health Care Opportunity 12. Performing clinical studies on anemia, immunity, wound infection, and wound healing as part of every space mission.

Mental Health Issues

The transition to long-duration space missions will require greater emphasis on ways to prevent and successfully manage an array of challenges to the cognitive capacities and emotional stabilities of astronauts who will find themselves in an isolated, confined, and hazardous environment. They will be devoid of much of what supports their emotional well-being on Earth and will need to develop and maintain new coping strategies appropriate to the unique environment of space beyond Earth orbit.

Current data on the psychiatric sequelae of long stays in surly environments come primarily from studies of military personnel on submarine duty, Antarctic field scientists, and Biosphere inhabitants (Billica, 2000), as well as more limited experience on the Russian space station *Mir*. These data suggest that the incidence of discernible psychiatric symptomatologies, including depression, anxiety, substance abuse, and psychosis, ranges from 3 to 13 percent per person per year, depending on the setting (see Tables 3-2 to 3-7). Transposed to a six- or seven-member space crew on a 3-year mission, the likelihood that psychiatric problems will arise on such an expedition is not insignificant but is less than 54 percent—(3 percent/year) × (six astronauts/year) × (a 3-year mission)—per astronaut during a 3-year mission among a space crew when one extrapolates from the crude available data on behavioral disturbances in space.

Such problems can range from simple boredom and fatigue to acute stress reactions, profound depression, and overt psychosis. Some mental health problems may become more likely over time as the cumulative effects of environmental and interpersonal stressors are magnified by the extended duration of the mission.

The NASA Experience to Date

Almost all of NASA's behavioral medicine experience with space travelers thus far has been with flights of relatively short duration (i.e., 2 to 3 weeks), where emergent signs and symptoms have included evidence of stress, anxiety, diminished concentration, depressed mood, malaise, and fatigue. These problems have been identified in less than 2 percent of astronauts, and their effects on individual and crew performance have reportedly been negligible (Flynn and Holland, 2000). As a result, with the exception of the astronaut selection process, the level of clinical and research interest in mental health problems that may affect human performance during space missions has been relatively low. At the same time, there is growing awareness that such problems could prove to be major impediments to the successful conduct of longer-duration missions.

Mental Health Aspects of Extended-Duration Spaceflight

Little is known about the psychological capacity of humans to withstand the stresses of long-duration space travel, but what is known (e.g., from the experience on *Mir*) is ominous. Experience with extended-duration flights, defined as flights longer than 100 days (about 1/10 the anticipated duration of a mission to Mars), suggests that boredom, fatigue, and circadian rhythm and sleep disturbances, coupled with the exacting human performance requirements of such missions, constitute risk factors for the development of depressive syndromes of various severities, anxiety and irritability, and at times, dysfunctional interpersonal relationships, either within the spacecraft or between astronauts and ground personnel. On missions beyond Earth orbit, in which spacecraft crews will be isolated and confined to a relatively small living space and in which medical evacuation will not be an option, the development of these and other mental health problems may exert cumulative detrimental effects both on individual astronauts and on their fellow crewmembers sufficient to jeopardize the mission.

Meeting this challenge will require a reassessment of the mental health needs of astronauts in the context of NASA's overall health care program. Areas of renewed emphasis and support should include premission psychiatric evaluation; intramission psychiatric support and treatment, including the possibility that acute interventions may be required, such as in a major psychotic break, possibly with the use of forcible restraint and psychoactive drugs; and a program of postmission assessment, follow-up, and intervention where appropriate, as discussed in Chapter 5. For international efforts involving multinational crews, language and cultural differences, along with different approaches to diagnosis and treatment, will complicate these tasks. Accordingly, the United States and its partners in space must move toward the development of a health care system for astronauts (see Chapter 7) with a common language, common diagnostic criteria, and common standards of care. In so doing, some suggested areas of emphasis (see also Chapter 5) are as defined below.

Health Care Opportunity 13. Developing methods for the identification and management of mood disorders and suicidal or homicidal ideation and developing protocols for the management of violent behavior, including crisis intervention, pharmacological restraint, and physical restraint.

Premission Screening and Selection

The issue of crew selection needs to be rethought in the context of long-duration space missions. As detailed in Chapter 5, valid and reliable psychiatric screening instruments should be further developed, tested, and refined. The process of astronaut selection into or out of the program must include efforts to predict crewmembers' responses to the anticipated stresses of long-duration space missions on the basis of data derived from studies carried out on the ISS as well as in analog environments. Assessment of interpersonal, leadership, and followership skills, problem-solving capabilities, and emotional stability under conditions of extended isolation are some of the areas appropriate for future research. In addition, personal and family history data coupled with laboratory testing may identify individuals at increased risk for mental disorders (e.g., depression) that may emerge over the course of long-duration space travel and may be included in the database on which crew selection and flight assignments are based.

The development of more sophisticated selection and deselection criteria is a first step, to be followed by specific individual and group training in behavioral self-assessment and the self-administration of countermeasures designed for a range of anticipated health problems. Training individuals to work successfully within a small group and to engage in productive and collaborative problem solving with ground crews should be part of this process. The relative value of such training and the efficiency of specific countermeasures should also be assessed in the context of a well-designed program of behavioral and psychosocial research. Such studies should be carried out in the course of extended stays in space and in appropriate simulated or analog environments. Finally, the selection and training of the members of ground crews who will support and direct long-duration missions should be parallel to and integrated with the selection and training of the astronauts.

Dealing with Intramission Mental Health Problems

As mental health problems arise, some will respond to countermeasures designed and tested during premission training, whereas others will require the intervention of crewmembers or ground personnel. In this context, there will need to be clearly defined duties and responsibilities for such personnel, as well as appropriate training. Evidence-based clinical protocols and treatment algorithms that are specifically adapted for the space environment will need to be developed and tested.

In addition to the availability of psychiatric expertise on the ground, the preventive approach to in-mission mental health care should include the prior development of supportive and therapeutic relationships between mental health clinicians, crewmembers, and crewmember families. In this context, finding ways of ameliorating the effects of prolonged communication delays between space and the ground should be a research priority. An onboard formulary that anticipates the range of psychiatric problems that may or will arise is also essential, as is research on the pharmacokinetics of current and future psychotropic medications in microgravity environments.

Technology offers promise for maintaining behavioral health when professional assistance is millions of miles away. One type of countermeasure is software designed to self-diagnose and relieve emotional symptoms before they become a psychiatric condition. The first of most famous of these was ELIZA (Weizenbaum, 1966, 1979). It mimicked a nondirective therapeutic dialogue. The second generation of algorithm-driven software packages is now available. One example is the Therapeutic Learning Program (Gould, 1989). Software-guided therapy, when coupled initially with individual training and clinical oversight, has produced relief of symptoms ranging from headaches to anxiety and depression. Although, for the most part, the gains are nowhere near those obtained from individual treatment, the benefits are far superior to no treatment at all. An excellent and balanced review of this subject is contained in Massachusetts Institute of Technology Professor Sherry Turkel's book, *Life on the Screen: Identity in the Age of the Internet* (Turkel, 1995). It is likely that later versions of these methods will be far more effective and could be adapted, with adequate training and clinical oversight, for use by astronauts on long-duration missions.

Postmission Mental Health Care

Although acute and chronic in-mission psychiatric problems may jeopardize mission success, severe, postmission mental health problems, if directly attributable to astronauts' participation in long-duration missions, could jeopardize the program itself. The NASA-sponsored longitudinal follow-up study of astronauts' health has not revealed any untoward psychiatric sequelae of participation in the space program, although the stress of reintegration and postflight adjustment has been noted. The unpredictable effects of mission-related physiological changes and potential exposure to radiation, coupled with the emotional stress of reintegration following a long-duration mission, make it imperative that a postmission program of psychiatric assessment and individual, peer, and family support as well as mechanisms for long-term peer and family follow-up support be developed.

Neurological Issues

Nervous system dysfunction and illness may occur as a result of physiological adaptive responses (both neural and systemic) to microgravity, as a consequence of problems that arise within the spacecraft, or as a result of external events or exposures. In considering the neurological illnesses or events that might occur during a mission to deep space, the timing, type, and severity of problems should be taken into account. A logical medical distinction is to consider neurological problems that affect either the central nervous system or the peripheral nervous system, or both. This somewhat arbitrary distinction has practical implications from a diagnostic and therapeutic perspective. Available data indicate that a fairly high incidence of minor neurological complaints may occur (Tables 3-2 and 3-3). Contingencies are also needed for catastrophic neurological events, including those that threaten astronauts' lives and the mission itself. NASA recognizes that the occurrence of certain severe life-threatening events can exceed the capacities of either the astronaut crew or ground control to intervene medically. This concept of acceptable risk may be different for a single one-time mission than for a multimission exploratory program.

Acute Central Nervous System Illnesses or Events

There are insufficient data on which to base sound estimates of either the incidence of various central nervous system problems or the extent to which various central nervous system problems might occur during a long-duration space mission. So far, no major neurological illnesses have been reported. However, reports from the U.S. and Russian manned space missions suggest that minor neurological problems are frequently encountered (Tables 3-2 and 3-3). These include headache and vestibular dysfunction, particularly upon the initial entry into microgravity. A serious problem upon the return to Earth is orthostasis, with its consequent effects on many bodily systems including the central nervous system.

Closed head injuries and spinal cord injuries are among the most serious neurological events that could occur during travel beyond Earth orbit. Management and treatment of individuals with severe closed head injuries would likely be beyond the capability of an astronaut crew unless dramatic new approaches to clinical management are developed. Thus, injuries that produce very low Glasgow coma scales by today's standards will probably result in death, as they frequently do under the best of circumstances in current state-of-the-art medical centers. However, consideration should be given to training in the management of individuals with less severe closed head injuries. Individuals with mild or moderate closed head injuries may survive but remain disabled because of residual neurological deficits. Management issues today include placement of burr holes for evacuation of subdural hematomas, feeding and airway control, spinal cord stabilization, and management of bowel and bladder functions and infections. Other events to consider include toxic exposures, decompression sickness (especially in connection with EVAs), cerebrovascular-like events, spinal injuries, exposure to radiation, and seizures.

The current neurological clinical research program at NASA, although extensive, does not appear to be well coordinated among the various research organizations and those that design and conduct flight operations. Detailed treatment contingencies based on the accumulated evidence base for the entire spectrum of neurological diseases should be developed. Such treatments should be continuously reviewed and updated to maintain state-of-the-art readiness.

Health Care Opportunity 14. Establishing a coordinated clinical research program that addresses the issues of neurological safety and care for astronauts during long-duration missions beyond Earth orbit.

Urinary Disorders

Genitourinary disease may present as an infection, obstruction, or malignancy. Many potential genitourinary problems will be identified through standard screening. Renal stone formation (expected in 0 to 5 percent of astronauts) secondary to bone calcium mobilization and excretion in the urine is a well-identified concern in microgravity environments. The genitourinary effects of microgravity also include changes in urodynamics (unknown incidence) and urinary hesitancy (reported seven times). Nephrolithiasis is a concern during extended stays in microgravity, as alterations in calcium metabolism and hydration status have previously been identified in this environment. Dehydration (incidence, 0.01 per 14 days on the space shuttle) is a recognized problem (Lane and Schoeller, 2000). Dehydration or significant changes in pH and increases in calcium and citrate levels increase the risk of renal stone formation. Preflight screening should include appropriate ultrasound evaluation for nephrolithiasis.

Urinary tract infections are common (and are more common in females) and are generally easy to treat with antibiotics. Prostatitis can be treated with antibiotics. Preflight screening for prostate cancer by determination of the prostate-specific antigen level in serum and other evaluations, according to today's standards, appears to be adequate, although future consideration may be given to preflight ultrasound or other developing noninvasive methods.

Countermeasures for genitourinary problems are primarily oriented toward the prevention of nephrolithiasis through adequate hydration. The recommended daily fluid intake is greater than 2.5 liters. A more than adequate water supply must be ensured so that crewmembers do not hesitate to drink adequate volumes of water to prevent the formation of renal calculi. As ultrasound devices become smaller, it is likely that an ultrasound device will be standard medical equipment for all long-duration space missions. This would make it possible and desirable to perform in-mission screening for nephrolithiasis (to identify those who require medication or increased levels of hydration to treat calculi). As countermeasures are developed for the problem of bone mineral density loss in microgravity, it must be ensured that the solutions do not result in increased rates of renal stone formation secondary to alterations in calcium metabolism.

Some of the health care opportunities that may be explored to increase the future effectiveness of managing risks to astronaut health during space travel have been described in this chapter and are listed in Box 3-6. This is a short list of current opportunities. It is neither a comprehensive list nor a list of priorities but is presented as a list of areas of research and development to be considered. New opportunities, including some that may take precedence, will develop in the future as the field of space medicine continues to evolve.

CONCLUSION AND RECOMMENDATION

Conclusion

Space travel is inherently hazardous. The risks to human health of long-duration missions beyond Earth orbit, if not solved, represent the great-

BOX 3-6
Health Care Opportunities in Space Medicine

1. Expanding, validating, and standardizing a modified physical examination, the microgravity examination technique, and including a technique for pelvic examination for use in microgravity.

2. Developing an easily accessible database for medications on the spacecraft, including dosage, indications, adverse effects, and anticipated changes in the pharmacokinetic profile in microgravity.

3. Developing an easily accessible hazardous materials manual for space travel to aid in the surveillance, detection, decontamination, and treatment of chemical exposures.

4. Monitoring and quantifying particulates on a continuing basis.

5. Examining the capability of microbial identification, control, and treatment during space travel.

6. Developing methods for noise cancellation or reduction.

7. Standardizing ergonomic practices on the basis of the human body's response to the microgravity environment.

8. Developing methods to measure human solar and cosmic radiation exposures and the means to prevent or mitigate their effects.

9. Providing a thorough cardiovascular evaluation similar to the premission evaluation at the cessation of space travel to provide useful data as part of the continuum of astronaut care and to aid in establishing an evidence base for cardiovascular disorders during space travel.

10. Developing a program for instruction in basic dental prophylaxis, the treatment of common dental emergencies such as gingivitis, tooth fracture, dental trauma, caries, and dental abscesses; and tooth extractions.

11. Studying the bioavailability and pharmacological function of exogenous hormone therapy during space travel and, as new medical therapies for gynecological surgical conditions evolve, testing of these therapies for use during space travel.

12. Performing clinical studies on anemia, immunity, wound infection, and wound healing as part of every space mission.

13. Developing methods for the identification and management of mood disorders and suicidal or homicidal ideation and developing protocols for the management of violent behavior, including crisis intervention, pharmacological restraint, and physical restraint.

14. Establishing a coordinated clinical research program that addresses the issues of neurological safety and care for astronauts during long-duration missions beyond Earth orbit.

est challenge to human exploration of deep space. The development of solutions is complicated by lack of a full understanding of the nature of the risks and their fundamental causes.

- *The unique environment of deep space presents challenges that are both qualitatively and quantitatively different from those encountered in Earth orbit. Risks are compounded by the impossibility of a timely return to Earth and of easy resupply and by the greatly altered communications with Earth.*
- *The success of short-duration missions may have led to misunderstanding of the true risks of space travel by the public. Public understanding is necessary both for support of long-duration missions and in the event of catastrophe.*

Recommendation

NASA should give increased priority to understanding, mitigating, and communicating to the public the health risks of long-duration missions beyond Earth orbit.

- **The process of understanding and mitigating health risks should be open and shared with the national and international general biomedical and health care research communities.**
- **The benefits and risks—including the possibility of catastrophic illness and death—of exploratory missions should be communicated clearly, both to astronauts and to the public.**

NOTES

Mission Specialist Jeffrey A. Hoffman and Payload Commander F. Story Musgrave during an EVA, i.e., a space walk, on December 8, 1993, from the space shuttle *Endeavor,* during STS-61. The space walk to deploy solar arrays on the Hubble Space Telescope lasted 7 hours and 21 minutes. NASA image.

4
Emergency and Continuing Care

Dreams are like stars: you choose them as your guides and following them, you reach your Destiny.

Inscription on the wall of the science laboratory Destiny revealed on the opening day for Destiny on the International Space Station, February 11, 2001, accompanied by signatures of those who assembled the module

One or more acute severe emergency medical events such as a traumatic injury, toxic exposure, or acute cardiopulmonary decompensation will probably occur on a long-duration space mission. Some will resolve or at least improve quickly with treatment, others may require continuing care, and others may require resuscitative measures. The decision to initiate resuscitation or determine end points is complex and must be based upon the best judgment of the medical command on the International Space Station (ISS) and future spacecraft. Factors to be considered include whether there are single or multiple incidents, the resources available to treat the patient, and the operational impact upon the crew of the potential loss or extended disability of the patient. Case reports from the Australian National Antarctic Research Expeditions experience (Pardoe, 1965; Priddy, 1985; Taylor and Gormly, 1997) suggest that ingenuity and determination can be used during the treatment of unusual situations, so restrictive guidelines for the withdrawal of care or for the provision of only supportive care should be avoided. Other factors to consider include resource utilization in the event of multiple illnesses or casualties, identification of specific roles for caregivers, anticipated end points of treatment, and communication with the ground crew to assist in prioritization of care.

Health Care Opportunity 15. Developing a resource-based medical triage system that contains guidelines for the management of individual and multiple casualties during space travel.

ANESTHESIA AND PAIN MANAGEMENT

During long-duration space missions, anesthesia and pain management may be required for unanticipated accidents (e.g., fractures of bones, lacerations, or blunt trauma), medical conditions (e.g., appendicitis or a perforated viscus), and possibly, cardiopulmonary resuscitation. This aspect of health care in microgravity presents major challenges. Inhaled volatile and gaseous anesthetics must be avoided because of leakage of subanesthetic doses in a controlled environment, limited knowledge of gas diffusion in zero gravity, and the necessity for endotracheal intubation.

The spacecraft poses many challenges that need to be resolved before anesthesia and pain management can be delivered safely. First, the limited amount of physical space available not only will dictate the use of smaller equipment but also may preclude the use of some instrumentation that is standard during anesthesia delivery on Earth. The second challenge will be the selection of an astronaut(s) who has the technical skills necessary to administer the various components of anesthesia.

Airway Management in Space

It will be essential to have on board spacecraft astronauts who have requisite training in airway management, along with strategies for retaining these skills. However, success in applying some of the most effective methods of airway management requires frequent and regular practice, in the opinion of members of the committee who as part of their professional activities have expertise in airway management, and it is not likely that an individual with such characteristics will be on board the spacecraft. Therefore, techniques that are based on the likelihood of success should be chosen, and these should be consistent with the skills of the individuals on board. The types of airway equipment and techniques required will partly be based on the type of surgery and trauma anticipated. Decisions regarding long-term airway care need to be made before such equipment and techniques are selected. Lastly, because airway management is usually required for successful cardiopulmonary resuscitation, the need for airway management during anesthesia should be coordinated with the need for cardiopulmonary resuscitation.

To minimize morbidity and mortality, two approaches should be followed. The first is to use an anesthetic technique that does not require endotracheal intubation. The second is to avoid the use of neuromuscular system-blocking drugs. Anesthesia with the latter can facilitate an overall view of the trachea during endotracheal intubation, but if the trachea cannot be intubated, the patient is left totally dependent on the clinician for adequate ventilation. Controlling ventilation in a paralyzed individual is often difficult (Caplan et al., 1993; Parmet et al., 1998).

Little is known about airway management in an atmosphere of microgravity. Studies have been performed in simulated microgravity in a submerged full-scale model of a space module with neutrally buoyant equipment and personnel. The ability to control the airway in a free-floating submerged manikin was attempted with four different devices. Endotracheal intubation was unsuccessful in 10 to 15 percent of the situations, whereas other airway devices were more successfully implemented (Keller et al., 2000). Although this was useful initial information, the study with a manikin in no way simulates clinical conditions and the mental pressure associated with the need to control an airway in a life-and-death emergency.

These studies emphasize the potential for serious morbidity when intensive airway management is needed for either anesthesia or resuscitation. These clinical problems could contribute to the need for long-term care on board the spacecraft.

Because the possibility of failure of endotracheal intubation is real, less demanding techniques should be considered, even if they do not protect the airway from vomiting and aspiration as well as endotracheal intubation does. Insertion of a laryngeal mask is technically much easier, and it provides an excellent airway. In the simulated atmosphere of microgravity, use of a laryngeal mask has been more successful (Keller et al., 2000).

Anesthetics

Although inhaled anesthetics, both volatile and gaseous, are extremely useful and helpful, they should not be considered for use in space travel at this time, as stated above. Until the problems associated with gas-liquid interfaces in microgravity can be solved, intravenously administered anesthetics should probably be used. Although such anesthetics can be given by bolus administration, constant infusion achieves more sustained levels of anesthetic in the blood and brain, but constant infusion would require appropriate infusion pumps. These are small, however, and should be readily accommodated on a spacecraft. Nevertheless, basic technical procedures, such as procedures for the insertion of a reliable intravenous catheter, need to be reviewed, modified, or redesigned where necessary and need to be practiced.

In general, the drugs used should be associated with a rapid recovery and, preferably, should have an antagonist. These drugs should not cause significant cardiorespiratory depression. Unfortunately, no ideal drug with all of these characteristics exists at present.

Regional Anesthesia

The major advantage of regional anesthesia is that it anesthetizes only that part of the body that needs to be anesthetized and does not have the disadvantages associated with anesthetizing the entire body. It also has the advantages of maintaining the patient awake and facilitating a rapid recovery and postoperative analgesia.

Of prime importance on a spacecraft is what should be done if the regional anesthetic turns out to be inadequate for surgery. The nerve block procedure can be repeated (with the added potential complication of nerve toxicity), or a general anesthesia can be induced. Direct nerve injury and hematoma must be considered from use of the nerve-blocking procedure, although those are extremely rare complications on Earth. The team must be prepared to deal with central nervous system excitation and possible convulsions with short- or intermediate-action local anesthetics (e.g., lidocaine) and cardiotoxicity with the longer-acting drugs (e.g., bupivicaine).

Despite numerous potential problems, crew medical officers could be trained in the delivery of regional anesthesia in an effective manner. After conducting proper research on the effects of microgravity on the diffusion and nerve toxicities of local anesthetics, these will probably be the agents of choice.

Health Care Opportunity 16. Developing an anesthetic approach associated with rapid and comfortable recovery using anesthetic drugs with short durations of action or for which there are antagonists.

SURGERY AND TRAUMA

In the most basic sense, surgery is controlled trauma. Successful management of both accidental trauma and the controlled trauma of surgery requires several elements working in concert. From the standpoint of the health care provider, control of pain (anesthesia), control of bleeding (homeostasis), and technical expertise are just the beginning. Because anesthesia and pain control are considered separately, this brief summary will concentrate on the remainder of the significant issues surrounding surgery and trauma.

Physiological Responses to Injury

The physiological alterations that accompany long-duration space missions may well affect homeostasis, wound healing, and resistance to infection, all of which are crucial for recovery from trauma or surgery. Any wound sets into motion a complex series of events under neurohumoral system control termed the "response to injury." The time course has been fairly well delineated in the normal Earth environment, and the routines of postoperative and postinjury care are based upon the assumptions implicit in this model of events (Greenfield, 1996). The initial studies upon which this model is based were performed in the 1930s and have been refined by innumerable clinical and laboratory observations over the intervening decades to derive assumptions about fluid and electrolyte balance, changes in hormone levels, metabolic changes, cytokine activation, and immunological alterations after injury. Essentially nothing is known about how this neurohumoral response to injury is modified by microgravity.

Although it may not be realistic to replicate these myriad physiological studies during the postoperative period, carefully designed experiments with simple outcomes measures (supplemented by highly selected physiological, histological, or metabolic studies) performed in a microgravity environment such as the ISS after surgery in animals might yield significant information of value.

Take one example: the response to hemorrhage and fluid resuscitation. Cardiovascular adaptations and changes associated with microgravity will probably alter the response to hemorrhage and fluid resuscitation. Without gravitational pooling, will fluid volumes required for resuscitation from a standard hemorrhage be the same, greater than, or less than those required on Earth? What fluids should be used, and at what volume? What (if any) blood substitutes should be used? Would the use of a counterpressure device (such as military antishock trousers) help? Or would the use of such a device prove detrimental or nonproductive in an environment lacking the capability to obtain definitive control of hemorrhage?

To stop bleeding and achieve homeostasis, the surgeon applies mechanical methods (ligatures, cautery, hemostatic clips) that are likely to be quite effective in microgravity (Campbell and Billica, 1992; Colvard et al., 1992; Campbell et al., 1993, 1996; Campbell, 1999). In addition, new methods are continually being devised; for example, an ultrasound-activated method of sealing blood vessels has been demonstrated to have significant advantages in minimal-access surgery, supplanting the traditional electrocautery for many applications (Gossot et al., 1999). Continued developments can be anticipated, as this is a major area of focus in civilian and military medical practice.

Similarly, new methods of wound closure are continually being devised as alternatives to sutures or staples. For example, both biological and artificial adhesives are under continual improvement (Quinn et al., 1998; UHSC, 1999). These may be supplemented, when necessary, by a surgical zipper applied in such a fashion as to pull the edges of the wound together. This provides mechanical support for a "sutureless" closure (Mulvihill, 1999) that is quick and painless and that may be more resistant to infection than the closures achieved by more conventional methods.

Surgical homeostasis requires not only mechanical control of bleeding but also complex interrelated mechanisms of blood coagulation and fibrinolysis. As an example of unexpected issues that may emerge when actual surgery is performed or major injuries are treated, Campbell and colleagues (1993) noted that venous bleeding appeared to increase during surgery performed during parabolic flight. They hypothesized that gravitational forces on tissues normally contribute to the collapse of veins and help staunch the flow of blood when surgery is performed on Earth. In microgravity, however, such forces on tissue are nonexistent and external compression must be supplied. Practical lessons such as those about venous bleeding are best learned by carefully designed studies with animals.

Although immediate control of bleeding in microgravity has been demonstrated to be achievable by conventional means, it is not known whether long-duration space missions may result in subtle alterations in blood coagulation activity (e.g., platelet dysfunction or hyperfibrinolysis) that might require additional hemostatic maneuvers or the use of topical hemostatic agents. Anecdotal experience from *Mir* suggests that minor cuts take longer to heal, but it is dangerous to generalize this to an assumption that wound healing may be prolonged. Incomplete healing has been observed for muscles and bones injured in rats before spaceflight in the Cosmos biosatellite series, and brief daily exposure to near-infrared light appears to promote epithelial wound healing (Baibekov et al., 1995; Korolev and Zagorskaia, 1996).

Of greater concern than soft-tissue healing is the probable impact of bone demineralization upon healing of fractures. Fractures may be casted, internally fixed, or externally fixed. Although each of these methods has its advantages, neither casting nor external fixation would seem to allow access to a pressure suit, should this be required. Better information about how bone heals in microgravity would be extremely helpful. Innovative methods such as percutaneous injection of bone adhesives or growth factors-growth promoters might be available (AAOS, 1996).

Adequate nutritional support has revolutionized the management of surgical illnesses ranging from burns through trauma and infection. The neurohumoral response described above increases metabolic demands by 15 to 25 percent for a fracture of a major long bone (e.g., the femur) and 50 percent for multiple trauma (Greenfield, 1996). Ileus commonly accompanies burns, trauma, and surgical illness and may be exacerbated by the hypomotility associated with microgravity (Harris et al., 1997). This may make it impossible to maintain caloric intake; hence, short-term nutritional support may be needed during recovery from an acute illness or trauma. Consideration should be given to ways in which this might be accomplished within the realistic constraints of the mission. A simple way to monitor the adequacy of nutritional support, for example, urinary nitrogen excretion as a measure of nitrogen balance, would be helpful if it could be validated in microgravity.

Surgical Skills and Training

A word about remote or telepresence surgery is in order. By this method of surgery, a surgeon in a remote location controls robotic instruments that perform the actual surgery. It has been demonstrated to be feasible and is being explored for some applications in remote locations on Earth. However, even the slight transmission delay involved in transcontinental distances has proved problematic (DiGioia et al., 1998; Satava and Jones, 1998). Thus, in a mission beyond Earth orbit, transmission blackouts and delays will certainly limit its usefulness. Therefore, the roundtrip time lag (40 minutes or so once the mission is near Mars) negates the use of telesurgery and even real-time telementoring at this time, even though these methods have proved feasible on Earth.

Robotic surgery has captured public interest and enthusiasm. At present, surgical robots operate under the direct control of a human surgeon, who is usually in the same room. Currently, they are dependent on gravity. The robotic assistant is used to scale down movement and eliminate unwanted motions, allowing smaller, more precise sutures to be placed (Satava and Jones, 1998; Borst, 2000). Although rapid developments in this field are anticipated (Campbell et al., 1996; Satava and Jones, 1998; Campbell, 1999, Maniscalco-Theberge and Elliott, 1999), the application of robotic assistance to long-duration space missions is uncertain. Combinations of ultrasound-based diagnosis with computer-assisted positioning of therapeutic instrumentation may, in the future, make robotics a valuable technology for space medicine. A combination of computer-guided, robotic, and human hands-on expertise—including crew medical officer training and proficiency in evolving, minimally invasive surgical techniques such as laparoscopy—will be necessary, backed by data in an onboard computer might provide the best solution for complex low-probability problems.

Surgical Equipment

What surgical equipment should be taken on a space mission beyond Earth orbit several decades from now? Therapeutic options will be limited by the available drugs, equipment, and supplies. Thus, the choice of surgical equipment and supplies and the choice of procedures that will be performed seem to be inextricably linked. Yet, with computer-assisted design and computer-assisted manipulation (CAD-CAM) technology, it is possible to fabricate small parts to order from specifications contained in an onboard database or transmitted from Earth (Heissler et al., 1998; Okumura et al., 1999). Thus, it is conceivable that a seldom used surgical instrument (or an implant needed to stabilize a fracture) could be fabricated as needed and then recycled when it was no longer needed. Such technology is already available, and future application to surgical parts and instruments is likely a cost-effective way to achieve a just-in-time solution to a major problem on a spacecraft: the lack of hospital supplies. CAD-CAM equipment may well be included in a Mars mission for other (nonmedical) purposes; the task would thus be ensuring that materials and specifications were consistent with medical applications as well. The availability of this kind of technology dramatically expands therapeutic possibilities and emphasizes the need for skills acquisition by space crews.

By using today's standard of care as a guide, basic diagnostic equipment would include a high-resolution imaging system, an ultrasound unit, or conceivably, a compact magnetic resonance imager. This allows the detection of abscesses, bleeding, and pneumothorax and the performance of image-guided interventions (see Management of Abscesses and Soft-Tissue Infections below). In current trauma practice, a focused ultrasound evaluation has proved to be rapid and reliable in ruling in or out certain kinds of injuries (Bode et al., 1999; McCarter et al., 2000). This is another area in which multiple civilian applications are driving a rapid pace of technological innovation that could be adapted for use in space medicine if it is validated in microgravity. Whatever system is chosen may well need to be tested on healthy human subjects in microgravity, as anatomic landmarks shift without the constraints of gravity.

Technical Aspects of Surgery

In a microgravity environment, special mechanical problems such as anchoring both the patient and the operating team, maintaining a sterile field, and controlling blood and body fluids can be anticipated (Kirkpatrick et al., 1997). The mechanics of both open and laparoscopic techniques have been explored during parabolic flight, and the feasibility of these techniques with suitable modifications has been demonstrated. During open surgery, control of body fluids with sponges and suction appears to be adequate if the patient has normal hemostatic capabilities (Campbell and Billica, 1992; Colvard et al., 1992; Campbell et al., 1993, 1996; Campbell, 1999). Special adjuncts such as transparent plastic shields to provide containment of arterial bleeding can easily be designed and tested either in parabolic flight or on the ISS.

Laparoscopic surgery is an attractive option that has been demonstrated to be feasible in a porcine model during parabolic flight (Campbell et al., 1996). Other forms of minimal-access surgery may similarly prove to be both feasible and advantageous. Because the incisions used are small, recovery is usually rapid and limitations of physical activities are minimized. Better containment of body fluids is an additional advantage when working in microgravity. Minimal-access surgical simulators are already available, and it may be particularly easy for those who already have excellent eye-hand coordination skills to learn the techniques.

The preceding information and examples illustrate the accelerating integration of engineering and biology. The committee encourages the National Aeronautics and Space Administration (NASA) to track these and other emerging opportunities closely to apply them to space medicine as the technology develops to meet the need.

Prevention of Infection

Accidental wounds (such as lacerations and open fractures) are considered contaminated and require therapeutic administration of antibiotics as well as judicious surgical management to prevent devastating infectious complications in the best of circumstances (Singer et al., 1997; Luchette et al., 2000). Infection is possible even after clean, elective surgery. This may be more likely when surgery is performed during a mission. Breaks in sterile technique may be more likely in microgravity, the microflora is altered among individuals in close confinement (i.e., individuals share microorganisms), and there may be alterations in immune function in astronauts. Prophylactic antibiotics have been shown to decrease the probability of wound infection and are commonly used in current surgical practice (McCuaig, 1992; Bold et al., 1998; Weigelt and Faro, 1998). Selection of antibiotics for prophylactic and therapeutic purposes should be based upon knowledge of the microfloras that colonize the skin and gastrointestinal tract, coupled with pharmacological considerations (e.g., frequency and ease of administration and potential toxicity).

Management of Common Surgical Emergencies

Although one can anticipate the most common surgical emergencies from demographic and analog-environment data, unanticipated events can and do occur. A subarachnoid hemorrhage was managed by physicians in Antarctica, despite inadequate facilities, through ingenuity and innovation (Pardoe, 1965). Whether such a heroic effort could be undertaken or should be undertaken in the context of a space mission is doubtful, but it serves to show the human drive to preserve life, even against all odds. It also demonstrated how improvisation led to a successful outcome. Although it is not possible to plan for all emergencies, it may be possible to facilitate improvisation by providing the basic materials.

Fairly good data on the incidence of likely surgical emergencies are available from demographic and analog environments (Lugg, 2000). Anticipated emergencies include acute appendicitis, abscesses, incarcerated hernias, and trauma. Acute appendicitis is of particular interest because it provides a clear example of a common surgical emergency that has been approached in various ways over the past decades in an effort to avoid the necessity of an untrained person performing an appendectomy in a remote environment. In 1979, studies in Antarctica documented an increased incidence of acute appendicitis (Lugg, 1979) and led to the recommendation that the team physician undergo prophylactic appendectomy. The spectrum of measures available to avoid performance of an appendectomy in a remote environment ranges from prophylactic appendectomy before the mission to treatment of appendicitis with antibiotics with later (interval) appendectomy and to the use of antibiotics supplemented by percutaneous drainage of any abscess that might form.

Management of Abscesses and Soft-Tissue Infections

Abscesses and other soft-tissue infections are treated with antibiotics and by drainage or debridement, as needed. Image-guided (probably ultrasound-guided) percutaneous drainage (Nakamoto and Haaga, 1995; Montgomery and Wilson, 1996; Shuler et al., 1996; Miller et al., 1997; Scott et al., 1997; Wroblicka and Kuligowska, 1998; D'Agostino, 1999) would seem to be an ideal method for the management of deep-seated abscesses (as well as some less common conditions such as acute cholecystitis refractory to antibiotics). This technology is well proven as a safe and effective alternative to surgical drainage under the proper conditions and with concurrent antibiotic therapy.

REHABILITATION FOR ASTRONAUTS ON LONG-DURATION MISSIONS

Few controlled studies have assessed the contributions that rehabilitation interventions make to the prevention of the loss of lean mass in prolonged microgravity. The information learned from existing studies with both humans and animals is inadequate (NASA, 1997). This is true for existing studies of both of preflight ground-based training and in-flight exercise. Similarly, little is known about whether preflight training or onboard exercise has a mitigating effect on the loss of bone mineral density or muscle mass. Exercise strategies and techniques have not been selected on the basis of the results of evidenced-based trials, and treatment protocols have not been developed (NASA, 2000a).

Techniques for improving muscle strength and aerobic capacity preflight have not been tested for efficacy or efficiency. In-flight exercise has been aerobic; and although it helps to maintain a sense of focus and achieve a psychologically stable state (Linenger, 2000), in at least Dr. Jerry Linenger's case on *Mir*, the level of bone mineral density loss reached 13 percent after 5 months. Modification of existing practice, such as concomitant use of electrical stimulation or electromagnetic field stimulation, might provide an impetus for bone growth (Liu, 1996). Galvanic stimulation has had a benefit in maintaining muscle mass in patients postoperatively when they are kept in a non-weight-bearing state (Oldham et al., 1995). The use of devices that create bending moments of force might be useful, but no data are available to support this hypothesis.

Significant literature exists to support the use of other rehabilitative techniques to improve performance and provide a sense of well-being. These include biofeedback, gated breathing, transcendental meditation, and yoga. Movement therapies and deep soft-tissue massage have not been tried as part of routine training or in-flight management techniques designed to reduce stress. Data about the effectiveness of these techniques suggest that they may help with anticipatory nausea and anxiety in medical settings and may assist with regulation of the autonomic nervous system (Shapiro and Schwartz, 1972; Schwartz, 1979; Fehr, 1996; Murphy, 1996; Agathon, 1998).

The needs for rehabilitation on board during protracted missions, be they ISS missions or missions beyond Earth orbit, must address the issues of maintenance of lean mass and bone mass and restoration of bone mineral density after trauma or disuse because of illness and maintenance of musculoskeletal system performance (both fine and gross motor).

Strengthening exercise by an isotonic approach with resistance has been the technique used on Earth to preserve and restore lean muscle mass. This has not been methodically used on board spacecraft, and equipment that facilitates resistive exercise activity has not been routinely sent into space. Most efforts have been spent in supporting aerobic training, using a treadmill for exercise, but few data on the effect of this training on aerobic capacity or cardiac function have been published. Restoration of musculoskeletal system function following trauma has not been investigated and should be studied systematically in the case of injury (i.e., fracture or soft-tissue damage) on board a spacecraft.

The role of leisure and recreational activity to combat boredom and maintain fine motor and gross motor skills has not been fully evaluated. Yoga, relaxation training, and leisure activity may promote psychological well-being and improvements in coping strategies, vigilance, and performance. Nonpharmacological treatments for pain may also play a role in promoting well-being and reducing anxiety. Treatments shown to be effective include modalities of heat and cold, transcutaneous stimulators, acupuncture, and acupressure. The techniques of administration are easily taught, and the equipment needed is readily available and inexpensive. No information is available about its usefulness in space.

Restoration of function upon the return to Earth's gravity has been nearly complete in most instances, although full restoration of the bone mineral density that has been lost seems refractory to current treatments, which have mainly been exercise based. Nonpharmacological techniques for the stimulation of bone repair—for example, the use of electrical stimulators or electromagnetic field generators—have not been used.

NASA recently convened a working group on astronaut physical training and rehabilitation research and included in it a representative of the Russian Space Agency's rehabilitation program. During a 2-day workshop at the Johnson Space Center (NASA, 2000a), NASA asked the group to develop recommendations for preflight physical training, postflight rehabilitation, and rehabilitation after injury. NASA also asked the group to review its plans for a new bioastronautics facility, which would include a rehabilitation unit with increased staffing and new equipment.

In its key finding, the working group cited ample evidence of deconditioned astronauts returning from lengthy stays on *Mir* with physical defects. It said that this is an occupational health issue that must be addressed by an active rehabilitation program. The working group cautioned, however, that the techniques used for physical conditioning and rehabilitation on Earth have yet to be shown to be successful in the care of astronauts in space and that clinical research is needed.

The working group recommended that NASA investigate the use of new drugs for the treatment of osteoporosis, assess the effects of intense in-flight exercise, evaluate the Russian Space Agency's program of preflight and in-flight vestibular training, and develop strength standards for hand and forearm muscles (important for astronauts participating in extravehicular activities). Other recommendations focused on ways to increase motivation, compliance, and team cohesiveness.

The committee encourages NASA to continue its efforts in this area of astronaut health. Research on the efficacies of rehabilitation interventions and their mechanisms of action in this population should be undertaken by using ISS and Earth environments that are designed to simulate the effects of microgravity. Special effort could be made to determine what equipment needs to be installed on the ground or in space to facilitate appropriate rehabilitative interventions for preservation of lean mass and bone mineral density, good psychosocial interactions, and other conditions requiring rehabilitative treatment during a long-duration space mission.

CATASTROPHIC ILLNESS, DEATH, AND END-OF-LIFE CONSIDERATIONS

Guidelines for withdrawal of care and the provision of assistance to survivors should be developed for long-duration space missions. Resuscitative end point guidelines have been developed on Earth after years of experience and ethical discussion and have been limited to a few situations. There are no clear criteria that can predict the futility of cardiopulmonary resuscitation accurately. Responsibility for termination of medical resuscitative efforts rests with the responsible clinician (AHA, 2000).

Health Care Opportunity 17. Creating guidelines for withdrawal of care in space and for dealing with the death of a crewmember from physiological and behavioral points of view.

Consequently, an essential part of preparing for prolonged space travel is an open discussion of the principles and practices of resuscitative and supportive care. Practice principles to be developed include establishment of the chain of command during an emergency medical situation or during mission-altering or mission-ending situations in which restorative care is impossible, methods for stabilization of the astronaut-patient for the return to Earth, and methods for pain control and psychological support for life-ending conditions in the space environment. A clearly developed process for deciding what to do and how to proceed in the unfortunate instance of the death of an astronaut must also be fully undertaken and accepted by the astronaut crew, their families, and ground personnel before embarking upon a long-duration mission beyond Earth's orbit. There is a great deal of general maritime and Naval experience with this problem and a long history of practice in war and peace that could provide useful guidance on what to do and not to do in developing space medicine policy in this area.

PERSONNEL AND OTHER HEALTH CARE RESOURCES

The committee believes that, according to current standards of practice, the medical community and the general public would expect a physician to be part of a crew during a period of extended space travel and strongly supports that concept. There are, however, hostile remote environments on Earth where medical care is routinely provided by an individual other than a physician (British Antarctic stations and U.S. submarines). Whether or not a physician should be an integral part of a long-duration space mission will be determined by a number of factors including crew number, the length of the mission, the duties of crewmembers, and the standard of health care expected during the mission.

If a physician is part of the crew, however, a number of needs must be met. Cross-training of several or possibly all crew members in selected skills is necessary in the event that there are multiple illnesses or injuries that need simultaneous care or in the event that the physician is disabled or becomes ill in flight. At least one nonphysician on the space crew should be trained to the emergency medical technician-paramedic level by today's standards. The physician should have broad training and possess the general skill sets required for the evaluation and treatment of major illnesses and injuries involving organ systems and the identification and treatment of environmental illness and injury. The physician should also be trained in the technical procedures that will most likely be needed in the space environment, where evacuation to Earth for treatment is not possible. Methods should be developed for skill maintenance, retraining in those psychomotor skills that are most likely to be affected by microgravity, and the use of instructional and communication technologies such as telemedicine and virtual reality. The special skills actually required for performance of surgery can be acquired, augmented, or practiced by using simulators and a hybrid technology that has been termed "cybersurgery" (Satava, 1997; DiGioia et al., 1998; Satava and Jones, 1998). Training and retraining in clinical decision-making skills, clinical problem solving, and decision making for multiple casualties or illnesses are also necessary.

Health Care Opportunity 18. Developing a mechanism for skill maintenance and retraining in psychomotor skills during long-duration space missions.

The most important resource for care is the physician and the other members of the health care team, who must rapidly synthesize clinical and cognitive information to provide a diagnostic and treatment plan for the patient, as communication with medical specialists on Earth may or may not be possible or effective when needed. It is estimated that communication with Earth or satellite stations will be available only 50 percent of the time and that there will be up to a 40-minute round-trip communication delay, depending on the spacecraft's proximity to Mars.

These estimates reinforce the necessity not only for independent action but also for a full spectrum of resources on the spacecraft. Diagnostic and therapeutic resources should be available, including computerized teaching resources and computerized references. A comprehensive drug database would be helpful. The crew should also be able to repair, maintain, and if necessary, fabricate equipment to perform diagnostic and therapeutic tasks adequately. Medical supplies that can be reused, reconstituted, or somehow replenished in the spacecraft environment should also be part of this future.

Development of a Space Medicine Catalog and Database

During the space mission, routine surveillance of health measures should be conducted and all episodes of illness and injury should be recorded. Systematic record keeping with a computerized database will be necessary so that data can be accessed periodically and also used for identification of diagnostic and therapeutic errors and continued health care planning. Sex and cultural differences should be identified as part of surveillance of health measures. A centralized catalogue for all peer-reviewed manuscripts, abstracts, reports, handouts, mission log summaries, and monographs related to space biomedical research, health care, and relevant studies in analog environments (Lugg and Shepanek, 1999) should be developed. Access should be maintained through a centralized database with contemporary medical information on personnel and equipment. All material should contain the date and the responsible author(s). This is important, because resources could well be available but not used because they are not accessible and because promising information regarding countermeasures or important biomedical associations could be missed because the material is unknown or unavailable. The databases on board the spacecraft could be continuously updated as newer scientific information becomes available.

The Australian National Antarctic Research Expeditions Health Register is one model for such a database (Sullivan et al., 1991; Sullivan and Gormley, 1999). The database system should include descriptions of the illnesses and injuries that occur in microgravity, track normal health status measures, assess the types of illnesses and injuries and risks of exposure, determine the effectiveness of exposure modifications, track procedures performed by individuals and by the group, track the therapeutic agents that have been administered, and have a system for the coding of each episode of illness and injury according to the International Classification of Diseases and emergency medical code groupings. These codes should be expanded to include codes for dental problems. New emergency medical codes will likely need to be developed for the microgravity environment. Standardized rates for reporting of incidents, such as the numbers of incidents per person-day, should be adopted so that rates from different missions can be compared.

Health Care Opportunity 19. Recording routine surveillance of health status measures, incidents of illnesses and injuries, and their treatments in a database with standardized rates of occurrence so that data between studies and missions can be compared.

Health Care Opportunity 20. Developing and maintaining a centralized catalogue of all written materials related to space and analog-environment biomedical research and experience according to current medical informatics standards.

The health care opportunities that are described in this chapter and that may be explored to increase the future effectiveness of managing risks to the health of astronauts during space travel are listed in Box 4-1. These should be evaluated in addition to the health care opportunities provided in Chapter 3. Further opportunities will accrue as the field of space medicine continues to develop. As health care opportunities during space missions are explored, they will contribute substantially to the evidence base for the development of the appropriate means of management of medical events considered to be within the scope of care for long-duration space travel.

CONCLUSION AND RECOMMENDATION

Conclusion

Exploratory missions with humans involve a high degree of human-machine interaction. The human factor will become more important as the durations of missions into deep space with humans increase and as

BOX 4-1
Health Care Opportunities in Space Medicine (continued from Chapter 3)

15. Developing a resource-based medical triage system that contains guidelines for the management of individual and multiple casualties during space travel.

16. Developing an anesthetic approach associated with rapid and comfortable recovery using anesthetic drugs with short durations of action or for which there are antagonists.

17. Creating guidelines for withdrawal of care in space and for dealing with the death of a crewmember from physiological and behavioral health points of view.

18. Developing a mechanism for skill maintenance and retraining in psychomotor skills during long-duration space missions.

19. Recording routine surveillance of health status measures, incidents of illnesses and injuries, and their treatments in a database with standardized rates of occurrence so that data between studies and missions can be compared.

20. Developing and maintaining a centralized catalogue of all written materials related to space and analog-environment biomedical research and experience according to current medical informatics standards.

the spacecraft crew functions more autonomously, adapts to unexpected situations, and makes real-time decisions.

- *NASA, because of its mission and history, has tended to be an insular organization dominated by traditional engineering. Because of the engineering problems associated with early space endeavors, the historical approach to solving problems has been that of engineering. Long-duration space travel will require a different approach, one requiring wider participation of those with expertise in divergent, emerging, and evolving fields. NASA has only recently begun to recognize this insufficiency and to reach out to communities, both domestic and international, to gain expertise on how to remedy it.*
- *Engineering and biology are increasingly integrated at NASA, and this integration will be of benefit to the flexibility and control of long-duration missions into deep space. NASA's structure does not, however, easily support the rapidly advancing integration of engineering and biology that is occurring throughout the engineering world outside NASA. NASA does not have a single entity that has authority over all aspects of astronaut health, health care, habitability, and safety that could facilitate integration of astronaut health and health care with engineering.*
- *The human being must be integrated into the space mission in the same way in which all other aspects of the mission are integrated. A comprehensive organizational and functional strategy is needed to coordinate engineering and human needs.*

Recommendation

NASA should accelerate integration of its engineering and health sciences cultures.

- **Human habitability should become a priority in the engineering aspects of the space mission, including the design of spacecraft.**
- **Investigators in engineering and biology should continue to explore together and embrace emerging technologies that incorporate appropriate advances in biotechnology, nanotechnology, space worthy medical devices, "smart" systems, medical informatics, information technology, and other areas to provide a safe and healthy environment for the space crew.**
- **More partnerships in this area of integration of engineering and health sciences should be made with industry, academic institutions, and agencies of the federal government.**

Mission Control, Johnson Space Center, Houston, Texas, during the early portion of space shuttle mission STS-95 on October 29, 1998, overlooking the Flight Director (FD) and the Spacecraft Communicator (CAPCOM) consoles. NASA image.

Dr. Paul Stoner (left) and Dr. Jeff Jones, on-duty flight surgeons in Houston's Mission Control Center during the April 24, 2000, launch attempt of space shuttle *Atlantis* during mission STS-101. NASA image.

5 Behavioral Health and Performance

If the planet is ever terraformed, it will be done by Human Beings whose permanent residence and planetary affiliation will be Mars. The Martians will be us.

Carl Sagan, *Cosmos* (1980, p. 135)

Advances in the exploration and habitation of space environments over the last half-century have set the stage for long-duration expeditions beyond Earth orbit. A common feature of these initiatives will be extended stays by groups of humans in extraterrestrial habitats. The success of such endeavors depends on the behavioral health and performance effectiveness of multinational microsocieties living and working continuously in confined, isolated, and hazardous environments for extended periods of time.

The increasing durations of space missions and human habitation in space environments over the past four decades have presented a range of biomedical and behavioral challenges. These challenges have been met with remarkable success, often under adverse circumstances. The so-called human element, however, remains the most complex component in the design of long-duration missions into space. The imperatives of behavioral health and performance effectiveness therefore present major challenges to missions that involve significant increases in the time spent and the distance traveled in space.

The organization of work and living conditions in space environments must be a primary consideration in the operational requirements for such bold endeavors. Closely related considerations focus on the Earth-bound support and recovery systems interacting with the astronauts and the role of astronaut screening, selection, and training.

ASTRONAUT PERFORMANCE AND GENERAL LIVING CONDITIONS

Background

Hundreds of humans have now participated in missions that required the occupancy of spacecraft vehicles or space stations for periods of up to several months or in some cases a year or more under generally adverse environmental and behavioral conditions. Living space is confined, food is restricted in quality and diversity, there is a lack of privacy, and facilities for personal hygiene are limited. The quality of the environment provided by artificial life-support systems, compounded by high noise levels and unpleasant odors, is hardly comparable to that on Earth. Weightlessness requires motor and perceptual readjustments under conditions in which disorientation and motion sickness are common, at least during the initial exposure to space (SSB and NRC, 1987). Social interactions are limited, and sexual activity is constrained. Only distant and remote communication with family and friends is possible. Workloads can be demanding and stressful, with the ever-present danger of a major life-threatening system failure. All of these restrictions occur under conditions that make no provision for escape, at least during missions beyond Earth orbit.

Current Practice and Knowledge Base

Despite these conditions, neither the National Aeronautics and Space Administration (NASA) nor other agencies of the international space community have documented behavioral health problems sufficient in magnitude to compromise mission objectives, although they may have occurred or on other occasions have nearly reached the point of compromising the mission. It has been reported (Burrows, 1998), however, that three Soviet missions were aborted for reasons partly psychological. Other evidence supports the assumption that emotional or social problems could develop, and over months and years, these problems could grow to the point at which they disable an astronaut and limit crew effectiveness. Such evidence comes from three sources (Box 5-1): (1) anecdotal reports from astronauts and cosmonauts, (2) summary information from NASA on the incidence of adverse behavioral health events during space missions, and (3) findings from analog environments such as crews aboard submarine patrols or wintering over in the Antarctic. The 1998 report, *A Strategy for Research in Space Biology and Medicine in the New Century* (SSB and NRC, 1998a), also indicates the importance of a number of psychosocial issues.

Performance decrements are usually transient, although behavioral data from more extended missions have not been easy to access. The validity and reliability of participant self-reports, anecdotal and otherwise, as well as those of official reports, have been difficult to both access and evaluate, at least in part because of the traditional reluctance of flight-qualified individuals to be forthcoming about such behavioral health events. Moreover, the extent to which available accounts are representative of astronauts' experiences in general is not easily determined. The available, evidence-based data from space missions are thus clearly insufficient for the committee to make an objective evaluation or projection of the behavioral health issues likely to be involved in long-duration space missions beyond Earth orbit.

However, the available database from analog settings such as undersea and polar environments (Box 5-1) may be informative, at least to some ex-

BOX 5-1
Evidence of Emotional or Social Problems on Short-Duration Missions

Astronaut and Cosmonaut Reports of Personal and Social Problems of Adaptation

The diaries of cosmonauts and astronauts who have spent long periods of time in space describe some of the personal and social problems of adaptation that can occur during long-duration space missions. For example, cosmonaut Valentine Lebedev (1988), who usually had a relatively sunny disposition, described his mood turning distinctly sour relatively early during his 211-day space mission on *Salyut 7*. By the 9th day he had a conflict with his fellow crewmember (p. 39). Just over 2 months into the mission he said, "My nerves were always on edge, I get jumpy at any minor irritation." About half way through the trip, his anxiety interfered with his sleep (p. 291). As with cosmonauts and astronauts alike, however, he kept his feelings to himself (p. 158).

BOX 5-1 Continued

U.S. astronaut Jerry Linenger, who spent nearly 5 months aboard *Mir* with two cosmonauts, wrote that he "was astounded at how much I had underestimated the strain of living cut off from the world in an otherworldly environment" (Linenger, 2000, p. 151). Even though he had done all he could do to prepare for the mission, he experienced profound feelings of isolation and confinement as well as alienation from his crewmates (p. 152).

Cosmonauts and astronauts alike agreed that the most challenging interpersonal problems were not among the crewmembers but, rather, were between the crewmembers in space and the mission controllers on the ground. Lebedev portrayed the largest proportion of his contacts with his ground control negatively (e.g., pp. 96, 102, and 164). For Linenger, because of communication difficulties and his perception of little support from below, his aggravation with ground controllers in both Russia and the United States reached the point that he stopped communicating with them after the first month (pp. 123–127). His feelings are summed up in the title of one of his chapters, "Cosmonauts, Da! Mission Control, Nyet!" (p. 118).

It is important to recognize that no recent U.S. or Russian space missions have failed because of behavioral health problems that led to diminished performance. Some astronauts have suffered from space sickness and headaches, anxiety, anger, and depression and have faced sudden, life-threatening emergencies. So far as is known, however, all of them performed their jobs, and the missions were largely successful. Some crews may have gotten along better than others, but they worked effectively as a team when they had to.

Adverse Behavioral Health Events During Space Missions

NASA has accumulated data on the incidence of adverse behavioral health events from postflight medical debriefings. Among 508 crewmembers who flew a total of 4,442.8 days on board 89 space shuttle missions between 1981 and 1989, 34 "behavioral signs and symptoms" were reported to the medical staff, the most common being anxiety and annoyance (Billica, 2000). This calculates to 0.11 per 14 person-days, or about 2.86 per person-year. Among the seven astronauts who flew on *Mir* from March 1995 to June 1998, however, there were only two reported psychiatric events, for a yearly occurrence of about 0.77 percent.

It could be that the more experienced astronauts on *Mir* had fewer behavioral health symptoms than some of the crewmembers on the space shuttle missions, but this has not been documented. Also, it is difficult to know how to interpret these discrepant findings since it is not known whether the same criteria were used to diagnose the conditions. Finally, because experienced astronauts as a group are reluctant to report physical and behavioral symptoms to NASA physicians, this figure may underestimate their true incidence (Collins, 1974, 1990).

Behavioral Problems in Analog Environments

Data from analog settings are instructive since those who have spent considerable periods of time in isolated, confined, and harsh, dangerous environments have confronted many of the external stresses common to long-duration space missions. Two examples are sailors on U.S. submarine patrols and groups wintering over in the Antarctic. Four studies of the incidence of psychiatric disorders severe enough to cause the loss of a day of work or medical evacuation among submariners found rates per man-year ranging from approximately 0.44 to 2.8 (Wilken, 1969; Tansey et al., 1979; Dlugos et al., 1995; Thomas et al., 2000).

Among men and women spending 6 months living together during an Antarctic winter, where evacuation was nearly impossible, the Australian National Antarctic Expeditions Health Register estimated the rate of mental disorders to be 2.3 percent. As with the experienced astronauts aboard *Mir*, the incidence of behavioral problems was dramatically less among seasoned veterans. Two hand-picked six-man crews trekked around the Lambert Glacier Basin in Antarctica for 100 days and recorded their emotional state twice a week. Although there was some variation in mood and occasional increased interpersonal tension, these remained within normal limits, and all completed the treks successfully (Wood et al., 1999).

Additional sources of information on behavioral problems in analog environments are provided by long-duration isolation work from hyperbaric chambers and Earth-bound simulators (e.g., the European Space agency's ISEMNI and EXEMSI projects, the Russian Space Agency's *Mir* simulator studies), and the NASA-funded crewmember and ground personnel interactions study that took place during the space shuttle-*Mir* program.

Relevance of Previous Reports to Long-Duration Space Missions

It is difficult to confidently predict from these anecdotal reports from astronauts and cosmonauts, previous space missions, or analog environments the likelihood of behavioral health problems among crews of highly selected, experienced, and well-trained astronauts on long-duration missions. The clearest difference is that the missions described above were all of shorter duration than a mission beyond Earth orbit will be. No information exists on the extent to which the individual distress and the interpersonal frictions depicted in the two autobiographies or other anecdotal accounts are representative of the problems that other space crews have had. The overwhelming majority of members of space crews have neither written nor spoken about the behavioral health problems that they might have experienced. Also, the previous space missions did not rate the severity of the symptoms, their durations, the degree of impairment, or whether the less experienced crewmembers, such as payload specialists, suffered more than the astronaut-pilots and mission specialists. The data from missions in analog environments have several obvious limitations as well. These include differences in the environments and the missions, crew characteristics, crew responsibilities, and levels of external support (Palinkas, 1990).

Yet, there is reason to consider the possibility that the stresses of long-duration space missions could cause problems in individual and interpersonal adaptation, which could threaten mental health and the success of the mission. Numbers of astronauts have reported psychiatric symptoms associated with shorter space missions. Also, the rates of occurrence of psychiatric problems among sailors on submarine patrols and those wintering over in the Antarctic were similar to the rates of occurrence of other medical problems such as cardiovascular, genitourinary, and ear, nose and throat problems (Tansey et al., 1979; Palinkas et al., 1998; Thomas et al., 2000). Just as NASA should prepare to care for these potential physical problems in astronauts, so too should it ready itself to look after the behavioral health problems of the astronauts on long-duration space missions.

tent, with regard to the consequences of prolonged isolation and confinement (Gunderson, 1974a,b; Wood et al., 1999). Although the relevance of studies conducted in analog environments to long-duration space missions may vary (Stuster, 1986), they do appear useful for examination of intrapersonal and interpersonal issues. Such untoward effects as lassitude, apathy, and hygienic neglect, for example, have been reported in studies of unstructured time of variable duration with polar crews in analog environments (Taylor, 1987).

Among the more systematic approaches to analysis of the data from studies conducted in analog environments is a reported content analysis of diaries from polar expedition leaders that ranked the salience of behavioral issues (Stuster, 1996). Of the 22 categories ranked, "group interaction" was found to be the most salient, followed by "outside communication," "workload," and "recreation/leisure." More structured studies with recruits from the Australian National Antarctic Research Expeditions have focused on the effects of group composition on the adaptation of individuals to a remote hazardous environment (Wood et al., 1999). The roles of group membership, the degree of isolation, and selected individual characteristics were examined in relation to self-evaluation measures. The reported results indicate that the predominant dimensions by which the participants determined how well they were functioning within the group included work, social life, and internalized emotional state, all of which appear to play an important role in how aspects of life in isolation and confinement are evaluated.

Requirements for Additional Knowledge

There is an overarching need to enhance the evidence base on the organization of general living conditions and performance requirements for small groups of humans in isolated and confined microsocieties. The objectives of this knowledge-seeking endeavor would be to specify the conditions under which effective work performance of the group can be generated and maintained within the context of productive and harmonious living arrangements that satisfy both individual and group needs for life support (clean air and water, an adequate food supply, waste management and recycling, lighting, adequate clean clothing, and communications) and general living conditions (medical care, sleep and rest, privacy, exercise, social interaction and sex, leisure and recreation, housekeeping and maintenance, education, remote contacts, and useful work).

Significant amounts of ground-based as well as specialized space mission research and development will be necessary to generate the knowledge essential to make adequate provision for these requirements for life support and general living conditions.

Engineering and Technological Interactions

To some extent, systematic studies are under way on specific subsets of the list of requirements for life support and general living conditions, particularly those for life support, which mainly require engineering or technological solutions. Over the past decade, for example, study groups of the National Research Council have been determining spacecraft maximum allowable concentrations of potentially toxic chemicals in the air and water supplies of spacecraft and habitats (NRC, 1992, 2000). The interdependent nutritional (Stein et al., 1999; Lane and Schoeller, 2000) and food supply system (Hunter, 1999) challenges of long-duration space missions have also been the foci of both intramural and extramural research and development support by NASA.

In the case of the food supply system, for example, food is not only a key habitability issue but also a biomedical issue, as well as an issue of engineering and systems design. Considering the duration of missions beyond Earth orbit, a bioregenerative life-support system (e.g., crop growth) would be more cost-effective than the physical-chemical regenerative systems (e.g., freeze-dried storage) now in use. Bioregenerative systems, however, require the growing and processing of crops in situ, treatment of food wastes, and preparation of daily meals, all within severe constraints for which space-compatible technologies have yet to be developed. Under such conditions, the behavioral health and habitability challenge resides in the development of a bioregenerative cuisine of nutritious and appealing dishes that chiefly make up a plant-based diet with a range of choices acceptable to a historically omnivorous population.

Biomedical and Behavioral Interactions

For the group of requirements for general living conditions, relevant intramural investigations (e.g., by NASA laboratories) and extramural investigations (e.g., by the National Space Biomedical Research Institute and university laboratories) have been undertaken with a focus on human performance effectiveness in the technologically rich spaceflight environment. The effects of changes in circadian rhythms and sleep patterns on behavioral health and performance, for example, continue to receive attention in both ground-based simulation studies (Kennaway and Van Dorp, 1991; Ross et al., 1995; Dinges and Van Dongen, 1999; Wright et al., 1999) and experiments conducted in the course of actual space missions (Monk et al., 1998; Czeisler and Wright, 1999).

The results of these ongoing sleep studies indicate that a strictly scheduled wake-sleep cycle with dim light levels comparable to those currently provided on space shuttle missions is sufficient to maintain entrainment of the human circadian pacemaker to the 24.0-hour day for most, but not all, study participants. Misalignment of the circadian rhythm results in disturbed sleep, impaired performance alertness, waking-hour melatonin secretion, and reduced levels of nocturnal secretion of growth hormone. The results of these studies suggest that during long-term missions beyond Earth orbit, the use of stronger synchronizers such as brighter lights will be necessary to entrain the longer-than-24-hour intrinsic circadian period of all humans to the 24.0-hour day and to other day lengths such as the 24.65-hour solar day of Mars.

Even with appropriate alignment of the circadian period, however, experience from previous space missions suggests the more likely emergence of restricted sleep patterns in the average range of 4 to 6 hours per day. With that restricted sleep comes the risk of development of cumulative homeostatic pressure across consecutive days of inadequate sleep during long-duration missions beyond Earth orbit. Thus, one objective of ongoing simulation studies is determination of the extent to which the duration of sleep per 24 hours and the use of combined "anchor sleep" plus "nap sleep" opportunities each day can prevent or attenuate the development of cumulative fatigue and performance deficits related to sleep deprivation.

Interactions in the General Living Environment

Behavioral health issues related to privacy and to leisure and recreational requirements have received some attention since the beginning of space missions that have involved multiperson crews in orbital flights of more than a few days' duration (Frazer, 1968; Kabanoff, 1980; Kelley and Kanas, 1994). The interaction between the structural-physical design and the personal-social organization of space environments is likely to be the most critical in leisure and recreational pursuits as well as those activities related to personal hygiene, exercise, housekeeping, and maintenance.

The historical background of human space travel reflects a predominantly empirical and generally successful approach to these matters. The research and development requirement, however, for evidence-based coordination between design engineering, habitability considerations, and behavioral health imperatives assumes ever-increasing importance with extended mission durations. Under the closed-loop conditions of such missions, for example, all waste products (including human excretions, expired gases, fluids, etc.) must be repeatedly recycled. The unique behavioral health factors involved in the toleration and acceptance of such environmental constraints must be investigated and determined. The extent to which behavioral self-management techniques (Beck and Emery, 1985; Beck, 1993; Barlow, 1996; Cautela and Ishaq, 1996) can be expected to provide effective coping procedures for such intrapersonal challenges will depend on the development and testing of individualized self-monitoring and self-assessment methodologies of demonstrated validity and reliability.

Group Interactions

Perhaps the matter of highest priority in the performance and general living conditions domain is the development of an evidence-based approach to the management of harmonious and productive, small, multinational groups who must live and work together in isolated, confined, and hazardous environments. Although an extensive literature on the functioning of small groups in both analog environments (Helmreich, 1973; Gunderson, 1974a,b; Vinograd, 1974; Palinkas, 1990; Harrison et al., 1991; Taylor, 1991; Weybrew, 1991; Palinkas et al., 1995; Stuster, 1996) and experimental settings (McGrath and Altman, 1966; Emurian et al., 1981; Brady, 1990; Brady and Anderson, 1991; Duffy, 1993; Guerin, 1994; McGrath, 1997) has been generated over the past half century, the available knowledge base is deficient with respect to long-duration missions beyond Earth orbit in several ways. For example, findings from group studies conducted in one setting are often not applicable to groups functioning under other environmental conditions. Empirical results are typically of such limited scope that they lack practical utility and generalizability.

The conditions under which such experimental observations or even observations from analog environments are made usually differ considerably from those encountered in operational spaceflight situations. The benefits and disadvantages of traditional approaches to the study of small-group dynamics have been well documented. When observations of the behaviors of small groups are made when the groups are in their natural habitat or in an analog environment, the generally ethnological monitoring and recording of both current and long-term events lack experimental rigor. On the other hand, data gathered on small groups in controlled experimental settings may demonstrate functional relations, but the analysis of progressive changes in external influences and the development of internal group equilibrium is often neglected (Williams, 1974; Brady, 1990; Guerin, 1994).

Viewed from the perspective of groups as small social systems operating in multifactorial behavior-specific settings that comprise specific physical situations involving people, the NASA experience over the past four decades has been successful. Mission objectives have been completed at the level of quality desired, and the behavioral interactions among crewmembers on individual flights have enhanced (or at least have not compromised) the viability of the group as a performing unit. Importantly, crewmembers have accomplished their mission objectives without significant deterioration of their individual behavioral or biomedical well-being. Against this background of generally effective group performance in the course of missions of up to a year or more in duration, the issues to be addressed now are how to promote performance effectiveness, group solidarity, and personal well-being in the course of long-duration space missions beyond Earth orbit. The enhancement of this essential knowledge base must begin by identifying those features of small social systems that foster the effectiveness of groups functioning semiautonomously over extended periods of time (Hackman, 1990, 1998).

First, there is a need for evidence-based information on the appropriate partitioning of authority between ground-based mission managers and the space crew in accomplishing clear, unambiguous, and engaging objectives that orient and motivate group members to achieve the overall goals (Radloff and Helmreich, 1973; Sandal et al., 1995). The key question is how legitimate authority can be constructive and empowering while setting directions without dictating procedural details. How can the needs of the astronauts for substantial latitude in developing, executing, monitoring, and managing their own performance strategies be best accommodated within the overall direction required? How can relations between astronauts and the authorities on the ground and the inevitable disputes that occur between them be managed in real time?

The second challenge in fostering performance effectiveness is to create a well-composed group engaged in well-structured tasks (Nelson, 1965; Gunderson, 1966a,b; Doll and Gunderson, 1970). Collective work productivity in the context of harmonious living conditions requires the right people who are correctly configured (i.e., individuals with personal and skill characteristics that fit the work and the setting) and who are properly trained. As subsequent sections of this chapter will confirm, the experience and knowledge base in NASA as well as in other public agencies related to personnel screening, selection, placement, and training is extensive. In the area of group composition and group training, however, there is a serious lack of evidence-based information on the requirements of long-duration space missions. There is also a need for research on how the development of semiautonomous, task-oriented groups occurs over time. To complement the available data on relevant individual tasks, an enhanced knowledge base on how group tasks that promote motivated and effective performance should be structured is also essential.

A third requirement for effective group functioning is a supportive physical and organizational context (Noy, 1987; Stuster, 1996). The first order of business here is the architectural and human factors considerations that determine the extent to which individuals and groups can live and work together comfortably and productively. The organizational factors to be considered include the performance consequences (i.e., reward system) for the group, the communication-information system (i.e., data access and technical assistance for online decisions), and material resources (i.e., for work execution). The evidence regarding the potency and nature of these factors is clear and compelling, but the existing knowledge base on the most effective delivery system for ensuring their availability to semiautonomous, remotely located groups is in need of considerable enhancement. Such seemingly mundane, contextual factors represent powerful determinants of group performance effectiveness and individual behavioral integrity. There is, however, little or no evidence-based information on how the operational delivery and organization of these factors can be configured to optimally support the group (see the section Support and Recovery Systems later in this chapter).

The availability of competent leadership is the fourth factor needed to understand and manage the performance effectiveness of small operational social systems. The history of leadership research, including results of investigative studies on the topic, indicates that an investigative initiative must be structured to generate information that is more evidence based, trustworthy, and useable than the information from past research (McGrath, 1984, 1997). An approach that first identifies the leadership functions important to group performance effectiveness and then examines how and by whom these functions are best carried out seems to be indicated.

An analytical approach might begin with an examination of the functions best performed by the organizational managers who establish flight crews, mission objectives, and schedules (i.e., condition-creating directive functions including design, selection, group formation, and training). Further analysis would focus on the functions best performed by ground control managers who work with flight crews online and in real time during a space mission (e.g., predominantly support and technologically directive functions including organizational adjustments to ensure access to resources). It is important that the available knowledge on ground-based leadership characteristics and training requirements be enhanced. Two functions are best performed by the flight crew commander operating internally as a member of the space crew: (1) fine-tuning the space crew as an operational social system and (2) coordinating the interaction between the space crew and the ground-based managers. Only after such an analysis is it possible to evaluate the required leadership functions.

Despite the existence of a massive literature on the topic of leadership (Nelson, 1964a,b, 1973; Gunderson, 1966a,b; Hackman and Walton, 1986; Ginnett, 1993; Hackman, 1993; Nicholas and Penwell, 1995; Keller, 1999; Taggar et al., 1999; Boal and Hooijberg, 2000; Judge and Bono, 2000; Silverthorne, 2001), there has been a serious neglect by NASA in general of such functional approaches. Research on how various leadership functions and responsibilities should be partitioned among group and individual participants on long-duration space missions beyond Earth orbit is needed, as such ambitious undertakings will involve many different leadership roles.

There is another view, perhaps best expressed by proponents of the theory of complex adaptive systems (e.g., Paul Plsek). In this view, one cannot possibly predict the performance of a complex adaptive system (in this case, a crew, a spacecraft, and a ground support team on a 3-year mission). To optimize the likelihood of success, the design work should focus on the establishment of a few simple rules for crew behavior and let the crew and its leader(s) learn their way through the problems encountered on the mission on the basis of a common objective (a successful, safe mission) and powerful personal motivation (survival) (Zimmerman et al., 1998; Lewin and Regine, 2000). It is nonetheless clear that the promotion of performance effectiveness as well as social and ecological stability for small groups involved in long-duration space missions beyond Earth orbit will require evidence-based technological developments through an approach at the most fundamental scientific level. The committee prefers application of the former "we can predict" approach because it is proactive, more easily amenable to inclusion in astronaut training and is based upon the previous and continuously accumulating evidence.

Within the context of such an enhanced database on the determinants of effectiveness of small groups, further investigations will be required to analyze potentially disruptive influences on harmonious and productive interactions among members of the space crew. Among these are cultural differences among members of multinational crews, differences among members of different professional and technical disciplines, issues related to the distribution of authority, and sexual interactions. In the case of sexual interactions, careful consideration must be given to living arrangements that accommodate this challenge to group cohesiveness.

Use of Pairs of Transport Vehicles for Small Groups Traveling Beyond Earth Orbit

Virtually all aspects of the onboard health care support system, and particularly those related to behavioral health and group performance, would be enhanced, in the committee's speculative opinion, by a long-duration space mission design that used a pair of transport vehicles. In addition to the motivational advantages of the friendly competition that would be generated under such conditions, the reassuring presence of an accompanying group to support the mission would confirm the availability of a nearby resource base. To the extent that interactions between flight groups could be established and maintained—including intermittent rendezvous during the outbound voyage—general living conditions would be enriched and potentially disruptive within-group issues would be attenuated. Cost-benefit analyses would take into consideration the modest incremental engineering and human systems requirements, including the support, training, and recovery requirements, over those for a single space transport vehicle beyond Earth orbit, which itself represents a major funding investment.

SUPPORT AND RECOVERY SYSTEMS

Background

Experiences over the past several decades in space and analog settings indicate the importance of a behavioral health role in supporting both participants and ground control personnel during and upon the return from extended space missions. Monitoring of both individual and group interactions has long been recognized as a potentially informative component of such exploratory or expeditionary endeavors, but the methods and procedures for such oversight have seldom been adequate for the task. Certainly, the behavioral health contribution to the planning and implementation of such support and intervention systems will need to be increased. As the time-distance dimension separating astronauts from their ground base increases, enhancement of behavioral health support systems will be required. Under such conditions, there will be a need for the development and the refinement not only of individual and group performance monitoring and assessment technologies but also of evidence-based behavioral interventions and effective countermeasures. Of at least equal importance is the role of behavioral health professionals in planning and implementing the reentry, recovery, and follow-up evaluation of astronauts returning from long-duration missions in space.

The present state of knowledge about support and recovery functions has mainly been derived from agencies responsible for the provisioning of various kinds of expeditionary forces (e.g., space crews, military teams, and labor groups such as those in Antarctica). Because of serious environmental hazards, uncertainties, and the need for minimal provisioning, expeditionary efforts have always been characterized initially by an authoritarian structure. Although it seems likely that space exploration will continue to be so characterized for some time, the frequent sequel to such expeditionary initiatives is envisioned to be the establishment of extended or even permanent settlements with increasing autonomy and increasing needs for new approaches to behavioral support.

Current Practice and Knowledge Base

The key element of NASA's space mission support systems is mission control. No other area of spaceflight operations has served as well or as long as mission control, which serves as NASA's institutional memory. Over the past four or more decades, mission control functions, which have been concerned with the monitoring of every aspect of every mission from the ballistic flights of monkeys Able and Baker to the current International Space Station (ISS) endeavor, have dominated virtually all spaceflight activities. It must be recognized that the record of safety and success that has characterized spaceflight is due to the support of mission control. The magnitude of this investment in support functions can be gauged by even casual observation of the number of individuals at Houston Flight Center computer stations during the continuous monitoring of every space mission. Perhaps the most striking contrasts are those that characterize the so-called manned missions, in which the number of Earth-based mission control personnel on duty can outnumber the inflight crew and passengers by more than 100 to 1.

From a behavioral health perspective, the communications functions of the inflight support systems are of the utmost importance. Aside from routine operational interactions, frequent communications with medical support personnel are available, along with less frequent opportunities for interactions with sources of emotional support (e.g., family, behavioral health personnel, flight surgeons, and others). Thus far there have been few provisions for ensuring confidentiality in these transactions, and there are indications that ongoing monitoring, at least of voice communications, may have some operational value for assessments of the behavioral status of in-flight crew members (Kanas, 1991). The Crew Status and Support Tracker questionnaire, completed in flight on a weekly basis by U.S. astronauts on the Russian *Mir* space station, has also been used for assessment of individuals (e.g., their mood, morale, level of privacy, physical status, social interactions, and other behavioral indicators). The validity and reliability of this assessment procedure, however, have yet to be determined.

Other features of the current support system appear to focus on behavioral health. For example, favorite foods and surprise presents are dispatched with periodic supply vehicle flights. Two-way communications with Earth-based family and friends are intermittently enhanced via combined audio and video. The transmission of computer-based picture albums of spouses, children, friends, and coworkers is enthusiastically endorsed by astronauts and cosmonauts. E-mail and ham radio capabilities are also available, along with private conferences with the NASA Behavioral Health and Performance staff. Recreational software, audiotapes, and videocassettes (and DVDs, now available on the ISS) are provided for leisure (Flynn and Holland, 2000). Over the past decade, debriefing protocols focusing on the well-being of the individual and on her or his reintegration with family and friends have been developed. Debriefing also identifies and helps to address any residual difficulties that may have developed within and between crew members, mission control, or family, friends, and coworkers. Currently, however, constrained access to debriefing data limits the extent to which these procedures can be adequately evaluated.

Requirements for Additional Knowledge

Ground Support and Space Mission Interactions

Perhaps the most important support system exists at the interface between ground crews and those in space. Disagreements between the "sent" and the "senders" have a long history. During World War II, submariners on 60- to 90-day patrols often complained of the unrealistic orders that were sent from rear-echelon officers thousands of miles away (O'Kane, 1977). More recently, anecdotal reports suggest friction between astronauts and ground crews, particularly during missions of relatively long duration (i.e., to the ISS). The impact of friction between the sent and the sender in both space and analog environments has been described by Nicholas (1987).

Friction develops because, originally, the senders were and have been exclusively responsible for the consequences of the activity, pay explicit fees, and demand absolute control over the performances of the sent. As groups are involved in longer stays in space and particularly beyond Earth orbit (or are transported to established settlements), however, the needs and aspirations of the sent become progressively more influential relative to the goals and objectives of the senders. Certainly, the increased time delay in the communication between space crews and ground personnel will require that astronauts themselves make some acute clinical judgments. Even a physician on board will not guarantee the availability of the expertise and capability needed to deal with some of the medical or behavioral issues that may arise.

This changing relationship between the senders and the sent is the fountainhead for the development of social structure and governmental policy as manifest in empire, colony, and emergent independent states. An enhanced knowledge base on how effective human support systems can be established and maintained as well as how internal and external factors influence their operational effectiveness needs to be developed.

Among the most important core functions of the ground-based support system is the maintenance of communication exchange capabilities between the astronauts and Earth-bound family, friends, and coworkers. Although these supportive interactions are both essential and heartily endorsed by all concerned, a host of unanswered questions regarding the nature of such exchanges remain (Palinkas, 1992; Kelley and Kanas, 1994). For example, family illness or death may develop during a long-duration space mission and information or the lack of such could significantly affect astronaut attitudes and behavior. The extent to which such activities must be enhanced or modified to provide behavioral health support for long-duration space missions remains to be determined and should be part of the behavioral research agenda on the ISS and other analog venues. Therefore, an evidence-based approach to the design of an effective behavioral health support delivery system for long-duration space missions beyond Earth orbit is needed. There is extensive military experience with this problem that may provide a useful guide to the consequences of various policies as they are being considered.

The interactive monitoring of long-duration space missions will depend on communication modalities and verbal interactions. The availability of automated interactive simulation technologies makes possible an experimental approach to determining the most effective mix of communication modalities. Technological research is needed to address constraints likely to be associated with the expected transmission time delays (with round-trip delay times estimated to be as long as 40 minutes).

Postflight Support and Reintegration

Despite the numerous unresolved issues described above, there have been no documented reports on the incidence among astronauts of psychiatric problems that are directly attributable to the spaceflight experience or that have compromised mission objectives. It is nonetheless reasonable to expect that physical changes like weakness, decreased immune system function, and decreased bone mineral density may add to returning astronauts' sense of vulnerability (Box 5-2). A significant period of readjustment, in-

BOX 5-2
Recovery and Reintegration

After 211 days in space aboard *Mir*, cosmonaut Valentine Lebedev ruminated while getting ready for landing, "We are anxious; who knows why? What's it like down there? We're no longer accustomed to life on the ground. Our lives are attuned to this small island in space, and suddenly here we come, back to the Big World. We don't feel comfortable with this idea" (Lebedev, 1988, p. 340).

Similarly, astronaut Jerry Linenger, reflecting on his nearly 5 months in space aboard *Mir* with two cosmonauts, wrote that he "was astonished at how much I had underestimated the strain of living cut off from the world" (Linenger, 2000, pp. 151–152). Even though he had done all he could do to prepare for the mission, he experienced profound feelings of isolation and confinement as well as alienation from his crewmates.

cluding medical and psychological follow-up, will be needed upon the return to Earth.

A follow-up study of individuals who wintered over in the Antarctic, for example, found a larger than expected incidence of postmission physical and psychological problems, including heart attacks and suicide (Palinkas, 2000a,b). Although these findings have not been replicated for other Antarctic crews or astronauts returning from long-duration missions, they highlight the importance of postmission evaluation of astronauts' physical and mental health coupled with support and intervention where necessary.

Monitoring Behavioral Health and Performance

The interpersonal problems that astronaut crews can endure for short periods may become intolerable over an extended period, thereby impairing crew performance. Schisms, friction, withdrawal, competitiveness, scapegoating, and other maladaptive group behaviors are found among highly competent men and women working together in normal terrestrial settings; they can also be expected among astronaut crews. NASA should anticipate interpersonal problems that may arise and rehearse solutions that can be used to maintain crew performance.

NASA spaceflight support system personnel have continued to research, develop, test, and apply environmental and biomedical monitoring technology (Mundt, 1999; Pollitt and Flynn, 1999; Schonfeld, 1999; Yost, 1999). Monitoring of behavioral health and performance, however, presents some special problems that have not been easily managed even in face-to-face settings, much less at the distances that separate Earth-bound support systems from the astronauts on a mission. A range of in-flight performance evaluation issues have yet to be resolved if adequate assessments of the effects of long-term space missions on human performance and behavioral health are to be provided.

In bridging this gap, early developments in the analysis of verbal interactions, both vocal and nonvocal, have shown promise of providing effective approaches. Over the past several decades, computerized methods for evaluating verbal communications have been refined as potential early-warning indicators of more general performance changes. One diagnostic technique (Gottschalk and Gleser, 1969) has been refined to reflect affective changes on a series of rating scales (e.g., anxiety, hostility, and depression). The method uses computerized scoring of speech transcriptions and is under evaluation for its validity and reliability. In addition, direct voice-analysis methods have already been used in spaceflight applications (Kanas, 1991). Continued refinements in computer technology will doubtless enhance the ability to discriminate vocal changes that are indicative of disruptive interactions that may have adverse effects on performance (Lieberman et al., 1995; Pickett et al., 1998; Wirth et al., 2000).

The feasibility of assessing the stability and precision of performance of individual crewmembers with computerized test batteries has now been demonstrated in the course of several space shuttle missions (Manzey and Lorenz, 1998; Monk et al., 1998; Brady et al., 1999; Eddy et al., 1999). For example, self-report rating scales, timing, learning, memory, and psychomotor components were included in a brief 20-minute performance battery scheduled recurrently during the 10-day STS-89 space shuttle mission (Brady et al., 1999). The demonstrated stability and maintained sensitivity of the indicated measures over extended time intervals confirmed their effectiveness in detecting even small-magnitude behavioral changes, that is, changes that occurred below the threshold of spaceflight duty performance decrements that would require intervention with a countermeasure. The Windows Space Flight Cognitive Assessment Tool (WinSCAT) is a computerized cognitive test battery that measures performance functions. It is similar to the instrument used in the STS-89 space shuttle study. The continued development of such technological approaches to the assessment of behavioral integrity and evaluation of the efficacies of countermeasures is essential not only to ensure the success of extended spaceflight missions but also to enhance safety and the quality of life in many applied settings (Kelly et al., 1998). The outcomes measures for these spaceflight studies were stored in onboard computers for postmission analysis. Further real-time evaluations will require online downlink capabilities in studies conducted during long-duration space missions.

At least two additional assessment instruments are under development. The Space Behavioral Assessment Tool (SBAT), is a measure of mood that consists of functional components that are being adapted for inflight repeated performance measures applications. The second assessment measurement under development is the Space Flight Fatigue Assessment Tool (SFAT), which is a measure of fatigue.

Psychophysiological monitoring could add an important dimension to the evaluation of the behavioral status of astronauts. Bioengineering initiatives will be required to develop more portable, noninvasive, and natural on-body instrumentation (e.g., nanotechnology) to record and downlink psychophysiological measures for the assessment of affective changes and cognitive dysfunctions of relevance to performance integrity (Cacioppo and Tassinary, 1990). Among the more obvious candidates for such psychophysiological monitoring would be heart rate variability, electrocardiogram waveform, pulse volume, facial muscle action, blink rate and magnitude, ear canal and skin surface temperatures, as well as multiple-electrode electroencephalogram measures.

Even though monitoring technologies have focused on individual performance and behavioral health, there is also a need for methods and procedures that can be used to evaluate group interaction patterns under spaceflight conditions. Standardized systems for the monitoring and evaluation of the interactions between the members of small groups are available and could be adapted for use with space-dwelling groups through the use of audio-video downlink capabilities. One such psychometrically robust instrument, Systematic Multiple Level Observation of Groups (SYMLOG) (Box 5-3), adapts easily to individuals of different sexes and cultures and has been demonstrated to be valid and reliable in both military and expeditionary operations (Bachman, 1988; Bales, 1999).

Countermeasure Development and Implementation

Substantial portions of the NASA research and development investment in life sciences have been and continue to be devoted to the development of countermeasures. In large part, the activities of the recently established Na-

BOX 5-3
Systematic Multiple Level Observation of Groups

Systematic Multiple Level Observation of Groups (SYMLOG) (Bales, 1999) is a method of group assessment. SYMLOG rates each group member on three continua: dominance versus submissiveness, friendliness versus unfriendliness, and being accepting of task orientation imposed by authority versus being nonaccepting. It also has scales for rating of the values as well as specific group behaviors. SYMLOG is psychometrically robust, has demonstrated validity, and is easily adapted to individuals of different sexes and cultures. It has been translated into 19 languages.

This method has been shown to be effective in distinguishing average from superior naval crews and officers on active duty in the Atlantic and Pacific fleets. In April and May 2000, SYMLOG was used by a Dutch army-sponsored climb of Pumo-Ri, a 7,200-meter mountain west of Mt. Everest. On the way to the top, the behavior of a group of nine climbers was rated daily by SYMLOG for evaluation of fitness and other performance factors.

tional Space Biomedical Research Institute, funded by NASA, are focused on this agenda. However, the core investigative activities—on radiation effects and cardiovascular, immunological, and neurovestibular functions, among others—do not reflect a strong emphasis on behavioral health issues. The one project on human performance is devoted almost exclusively to the sleep studies cited in the Astronaut Performance and General Living Conditions section of this chapter.

Some of the most important behavioral health and performance countermeasures are incorporated into NASA's screening, selection, and training procedures. The demonstrated effectiveness of these long-practiced methodologies in selecting and preparing candidates for participation in spaceflight operations has served NASA well by minimizing the number of occasions on which interventions with countermeasures were required.

There continues to be a need, however, for attention to potentially disruptive personal interactions among the members of groups of astronauts on space missions and between astronauts and Earth-bound support groups. Although investigations of countermeasures will necessarily focus on communication modalities and patterns, an experimental analysis of the most effective coaching functions and technical support interventions will likely yield useful results. It is also important that countermeasures be differentially adaptable for individual and group interventions. Other examples of psychosocial areas needing more attention include asthenia and other psychiatric syndromes (alluded to above), the displacement of negative emotions from crewmembers to mission control personnel, phasic changes in crew cohesion and mood over time (e.g., the "third-quarter phenomenon"), the importance of different leadership roles at different times during a mission, the effects of crew size and minority status on cohesion, cultural and common-language issues, the utility of voice analysis techniques in the monitoring of crewmember tension, and the need to reinforce positive attributes (e.g., leisure time interests, self-esteem, and creativity).

Recovery and Debriefing

At present, medical follow-up of returning astronauts and cosmonauts continues as long as contact can be maintained with the individuals involved. Ostensibly, the tracking procedures are in the process of being formalized into a more comprehensive data collection and management plan. If these initiatives incorporate behavioral health factors, a rich source of space mission debriefing and postmission (recovery) data can be established for further behavioral research. Attention can then be directed toward evaluation of the validity, reliability, and effectiveness of debriefing and follow-up procedures aimed at facilitating successful reintegration and readjustment to family, friends, and Earth-bound living conditions. The availability of such a longitudinal set of data could also support research to improve intra- and interpersonal support and countermeasure interventions during space missions.

Postmission debriefings and longitudinal behavioral health monitoring also provide the opportunity to evaluate the long-term effectiveness of premission training and intramission behavioral interventions, as well as to collect data on behavioral, social, and cultural issues that may not have been obvious during the premission and intramission phases. The relative importance and the long-term effects of individual and group factors and the availability of individual and group support can also be assessed during a prolonged period of follow-up.

After long-duration missions astronauts should participate in a structured recovery program with both physical and behavioral health components. Reintegration challenges include redefinition of astronauts' roles and relationships within NASA as well as with their families and communities. These processes will be affected by behavioral, cultural, and organizational factors and the abilities of the sent and the senders to realign their expectations of each other. Systematic study of the recovery process should be part of NASA's research agenda to guide the care of future generations of space explorers.

SCREENING, SELECTION, AND TRAINING

Background

Procedures for screening, selection, and training of flight personnel have a long and distinguished history among industrial nations with advanced military defense and air transport capabilities. Since its inception more than four decades ago, NASA has enjoyed ready access to the resources of the nation's military, including access to standardized and well-validated methodologies for screening, selection, and training. Beginning as early as the Project Mercury initiatives of the 1960s, a remarkable record of accomplishment has characterized a NASA space program that owes its success in no small measure to the effectiveness of these discriminating procedures. Moreover, similar programmatic approaches to screening, selection, and training have been adopted by such ISS partners as Russia, Germany, France, and Japan (Santy, 1994; Holland, 2000).

Current Practice and Knowledge Base

Screening and Selection

As the NASA space program has grown and prospered during the past half century, well over 1,000 candidates for astronaut and associated assignments have been interviewed and tested, with some 350 or more of the selectees having participated in space missions. The current methodology includes a detailed psychiatric interview that targets particular symptomatic conditions as defined by diagnostic criteria presented in the *Diagnostic and Statistical Manual*, 4th edition (DSM-IV) (Flynn and Holland, 2000). In addition, a test battery is administered that includes questionnaires with open-ended and multiple-choice questions, focused personality scales, and a computerized cognitive functions evaluation (Holland, 1997).

Apparently, none of the data from these intensive and extensive screening procedures are used in selection determinations except to "select out" candidates. Responsibility for "select-in" decisions resides exclusively with the Astronaut Selection Board. The extent to which the initial interview and test data are considered in the actual astronaut selection remains unclear. There is also little indication that the extensive screening interview and test data have been systematically analyzed for their procedural validities and reliabilities or even collated to examine interrelationships between measures.

Regardless of its shortcomings, the process has made important contributions to the successful accomplishment of mission objectives, insofar as participants on space missions have remained free of serious behavioral disorders, at least for relatively short-term space missions of up to a year or more. Under the present circumstances, however, there is no way of determining whether the procedures currently in place will be adequate or even useful for the screening and selection of candidates for long-duration space missions beyond Earth orbit

The existing knowledge base is enhanced to some extent by simulation studies that have been undertaken in polar regions, which serve as analog environments for long-duration space missions. In a study with some 600 American men spending an austral winter in Antarctica, for example, pretest and intake interview information was found to be useful in accounting, at least in part, for the variance in individual performance measures (Palinkas et al., 2000). In addition, recent initiatives implemented under the auspices of NASA have involved the exposure of small groups of astronaut trainees to extreme polar environments for relatively brief intervals.

Although studies conducted in analog environments have long been a rich source of experiential data, there is little evidence that the systematic data collected from studies in such settings are relevant to the behavioral challenges of long-duration spaceflight. However, these simulations in analog environments do provide an opportunity to pretest and refine instruments and procedures that can be used to select individuals and to select for effective group interactions and performances. These first steps can lay the foundation for an evidence-based approach to the establishment of valid and reliable procedures for the screening and selection of individuals and small groups who will be living and working in confined microsocieties under isolated and hazardous spaceflight conditions.

Training

The training of NASA astronauts and other flight and ground support personnel is notoriously time-consuming, thorough, and highly technical. For the most part it has been effective in guaranteeing the successful accomplishment of well-defined mission objectives. To a considerable extent, the success of the training is attributable to the involvement of well-educated, very experienced, and highly motivated selectees who are in exceptionally good health and who have a level of intelligence that is well above average.

The focus of NASA's training objectives to date has been the mastery of skills required to operate advanced spacecraft technology in the course of relatively brief, highly choreographed missions within Earth orbit. Generally, these space mission assignments do not exceed a few weeks, although in selected instances several months may be involved. On such brief assignments, with skillful and highly trained crewmembers, individual behavioral health issues are of minor concern. Although interpersonal consultation and coaching resources are apparently available to trainees, behavioral health does not appear to be part of the formal curriculum, and access is provided only by voluntary request.

Until recently, group training appears to have been introduced into the mission preparation process only after a mission has been scheduled and a space crew has been designated. Considerable lead time (often a year or more) is usually involved in group training, and the ensuing team training process may involve not only the designated crew and supernumerary backups but ground support personnel and monitors as well. Like individual training, group training also appears to focus on the operational aspects of the mission objectives, with little programmed attention to group functioning per se. An individual and group training program is under development, however, with candidates for long-duration missions beyond Earth orbit specially selected as participants. The extent to which this early initiative will take advantage of the availability of significant behavioral health and performance inputs remains unclear.

Requirements for Additional Knowledge

Screening and Selection

The development of effective screening, selection, and training procedures must be driven by performance requirements and by the general living and support system conditions under which long-duration space missions will take place.

The first task is a systematic analysis of the validity and reliability of screening and selection procedures as they are currently practiced, beginning with a descriptive collation of the data and an evaluation of the interrelationships between the measures. Access and review of available outcomes-related data may also shed some light on the utilities of current procedures. At the very least, such an undertaking would set the stage for an evidence-based approach to screening and selection for long-duration space missions.

A range of newly emerging approaches to the study of intellectual aptitudes and personal traits should be explored for their predictive value in assessing levels of intellectual functioning in candidates for spaceflight. Although current screening approaches appear to involve only limited evaluation of cognitive function, intelligence and aptitude measures have been shown to predict performance more accurately than personality assessments. Specific constellations of abilities including a variety of aptitudes identified as "fluid intelligence" (analytical, creative, and practical intelligence) have been shown to permit individuals to solve novel tasks without using crystallized knowledge involving general information or vocabulary or previously developed problem-solving skills (Sternberg, 2000). Other measures of cognitive function have been shown to predict pilot performances related to information-processing speed and working memory (Salthouse, 1991; Hartley, 1999) as well as multiple-task attention, breaking-set, visual reasoning, and mental rotation skills (Kane and Kay, 1992). The relevance of these advances in cognitive function assessment to the enhancement of astronaut screening, selection, and training as well as intramission monitoring and countermeasure development resides in the results of recent studies conducted during both simulated and actual spaceflights (Manzey et al., 2000). Under such conditions, degradation in manual tracking ability and phasic decrements in cognitive function performance have been reported as a result of fatigue and other factors.

The utility of measures that provide an evaluation of personal traits as they may be related to successful participation in long-duration space missions will also require careful examination. Data obtained from actual and simulated space missions, as well as submarine missions, led to the conclusion that individuals strongly motivated for achievement adapted to these environments better than their peers did (Sandal et al., 1999). By contrast, studies of crews that wintered over in Antarctica found that ambitious individuals had lower ratings on some performance measures than peers who had more modest achievement needs (Wood et al., 1999; Palinkas, 2000a,b). Santy (1994) has also reported that selection procedures in other national space programs (e.g., those of Russia, Germany, and Japan) tend to favor candidates for longer-term space missions whose personality characteristics measure more in the middle of the motivational range.

A number of promising new developments in the rapidly advancing fields of neuroscience and molecular genetics hold considerable long-range potential for the assessment, evaluation, and training of future astronauts. Neuroimaging procedures involving functional brain scans are continuing to advance knowledge of the relationship between behavioral processes involving both cognitive and emotional interactions on the one hand and well-specified neural systems and regional structures on the other (Damasio, 1994; Kosslyn and Koenig, 1995; Cahill et al., 1996; Alkire et al., 1998). There is reason to expect that future developments in the functional imaging field will provide noninvasive methodologies that could enhance the feasibility of both training and behavioral health monitoring applications. These probable breakthroughs could be of significant use to NASA. Four brief examples of how NASA might use these advances for screening, selection, and training are described in Box 5-4.

It is also conceivable that the future of long-duration travel beyond Earth orbit will be significantly affected by the rapidly advancing field of molecular medicine (see Chapter 4), to the extent that its neuroscience dimensions could differentially reflect DNA variations of relevance to adaptation to stressful behavioral and environmental situations. It is important that the

BOX 5-4
Potential Uses of Neuroimaging Methods for Astronaut Selection, Training, and Intervention

1. Functional brain scans for cognitive function assessment could considerably enhance the understanding of an astronaut candidate's neurological potential for handling the mental training requirements. Positron emission tomography scans find that areas of the brain illuminate when certain mental tasks related to astronaut performance are performed, with the occipital lobe lighting up when certain visuospatial tasks are performed or portions of the frontal and parietal lobe being activated during mental rotation tasks (Kosslyn and Koenig, 1995, pp. 133, 149). One screening scenario might include the simultaneous administration of a test of a particular aptitude (e.g., mental rotation) along with a functional brain scan that would show the extent to which particular portions of the brain were illuminated during the period that the problems were being solved.

2. Functional imaging techniques show promise as means of assessing and predicting potentially disruptive mood states and of perhaps monitoring and intervening to soothe potentially disruptive mood states (Damasio, 1994). Present work illustrating those portions of the brain that are differentially associated with strong feelings of sadness and happiness or of fear and anger could well lead to more accurate ways of assessing characteristics that could be used to select out or select in an astronaut, as well as whether an individual has a predisposition to mental illness.

3. Neuroimaging also may have a role in astronaut training. An example might be the use of scans as an adjunct to assessments of cognitive and affective states during simulations to assist in countermeasure development.

4. Imaging may also be useful for inflight monitoring and assessment of changes in brain function from the baseline function as a result of stress, fatigue, or possible exposure to microdoses of toxins.

procedures for the screening and selection of astronauts involved in long-duration missions beyond Earth orbit be capable of validly and reliably discriminating effective group interaction skills and competences. Essential developments in this regard will depend on the availability of an expanded knowledge base on the requirements for harmonious and productive group functioning in the unique environment of space.

Training

Astronauts on long-duration space missions will confront a range of intra- and interpersonal challenges, the nature of which cannot be accurately determined at present. Therefore, substantive features of training must be based on continuously accumulating experiences in actual spaceflight environments and analog settings with specially designed algorithm software packaging technologies (Lipsey, 1993; Newman, 1997). Naturalistic studies on the efficacies of specific training procedures must follow in both simulated and actual space mission settings.

Personalized individual training approaches must also incorporate and evaluate countermeasures based on procedures for evaluation of cognitive and behavioral functioning that are adaptable for computerized administration as self-assessment and supportive intervention procedures (Wolpe, 1958; Beck and Emery, 1985; Power et al., 1990; Beck, 1993; Barlow, 1996; Cautela and Ishaq, 1996; Rosen and Schulkin, 1998; Lazarus, 2000). These programs have been designed within a stress management context and have been effective when combined with a range of interventions including biofeedback, relaxation techniques, systematic desensitization, and pharmacological treatments.

Empirical observations about the nature of both individual and group behaviors and about how behavior patterns influence performance effectiveness can guide decisions about group composition and training. Training approaches can build on experience gained in simulated flight exercises going as far back as World War II (Office of Strategic Services, 1948) and, more recently, on that gained in the Cockpit Resource Management programs used by airline crews. The development of strategies for conflict resolution should be explored as well (Fisher et al., 1994; Heifetz, 1998).

Among the more recent and relevant developments with respect to training for small group performance effectiveness is the distributed interactive simulation methodology. This simulation approach (Box 5-5) uses multiperson computer-generated workstation networks for selection and training in realistic environments (Pratt et al., 1997; Gillis and Hursh, 1999). This method permits the objective recording and evaluation of interpersonal interactions within and between small training groups under conditions that simulate long-duration spaceflights, as well as related ground-based monitoring and support systems.

Training for long-duration space missions must involve an integrated approach that includes ground-based monitoring and support groups specifically selected to participate in such operations. NASA behavioral health personnel should be directly involved in crew selection and in training crewmembers and ground control personnel in crisis intervention and problems with interpersonal functioning. In addition, appropriate assessment tools and countermeasure development will be required to address emergencies and technical assistance requirements under conditions that involve multinational crews and the complexities related to cultural and language

BOX 5-5
Distributed Interactive Simulation

Distributed interactive simulation environments have been developed and are based on multiperson computer-generated workstation networks that represent operational elements consisting of both individuals and functional groups. Such techniques involving real people can be used for selection and training under conditions of simulated mission operations in a realistic environment. Participants communicate via electronic channels to exchange information, discuss work requirements, and evaluate data for decision making; exchange the outcomes of specific actions; and evaluate mission-oriented scenarios. Space mission simulations also permit inquiries of Earth-based mission control for information or instructions, or both.

Groups of individuals are trained to interact within the simulation environment for the purpose of engaging with assigned crewmembers and Earth-based mission control. Distributed interactive simulation methodologies with performance tasks requiring repeated exchange of information among participants and between groups provide an automated means for the systematic monitoring and analysis of the effects of experimental variations on psychosocial interactions, decision making, and both individual and group performance effectiveness. The operational performance measures evaluated include pattern analysis, task completion, and timing parameters.

differences as well as under conditions that involve crews of mixed sexes and with command structure constraints (Kelley and Kanas, 1992; Holland et al., 1993). It is not enough to have the leader be the buffer, because the leader could be addressing specific problems or could be too involved in a task-oriented emergency. Finally, crew resource management for long-duration missions also requires consideration of technical as well as nontechnical skills (e.g., corporate citizenship, interpersonal skills, and compatibility). Individual differences in personality functioning become important when the job requires corporate citizenship or the use of "people skills" (Borman et al., 1997; Hogan et al., 1998; Mount et al., 1998; Salgado, 1998).

STRATEGIC RESEARCH CONSIDERATIONS

The conceptual and methodological challenges associated with designing, establishing, and maintaining functional systems that promote performance effectiveness and social and ecological stability for small groups involved in long-term space missions beyond Earth orbit will need to be approached at the most fundamental scientific level. Evidence-based technological developments can be facilitated by research methodologies that incorporate studies in analog settings and simulations of the environmental conditions and behavioral interactions that will exist in space over long durations. The approach should be explicitly experimental and should be dictated by both scientific and pragmatic considerations closely approximating procedures of established effectiveness in other areas of natural science (Biglan, 1995; Lattal and Perone, 1998). Without such a database of experimentally derived data, the overextrapolation of proposals for ecological systems and recommendations for space habitat designs (Sells and Gunderson, 1972; Singer and Vann, 1975; Maruyama, 1978) renders them incapable of ensuring the successful establishment of functional and enduring space habitats.

The relevance of such an experimentally derived knowledge base to the success of future space ventures depends in large part on the precision with which required performances can be specified and on how effectively they can be occasioned and maintained in individuals and groups. Despite the abundance of extensively reviewed experiments with small groups conducted over the past several decades (McGrath, 1984, 1997), the technology needed to analyze multiple modes of communication among and between groups and group members has not been available. With the advent of computer-based communications technologies (Duffy, 1993), the interactions of members of small groups and decision making by small groups can now be automated in the context of distributed performance sites. An empirical research base has been established to evaluate the effectiveness (Pinsonneault and Kraemer, 1990) and test the assumptions (Kraemer and King, 1988; Duffy, 1993) that underlie these technological developments.

Distributed communication interactions within and between space-dwelling and Earth-bound groups will be an inherent feature of long-duration space missions beyond Earth orbit. The availability of automated group interaction technologies not only opens the door to the enhanced precision of performance measurements but also provides a research approach to determining the correct mix of group interaction-communication modalities that maximize efficiency without overcomplicating the design of the system. Research applications of such computerized distributed interactive simulation systems can permit the specification, with nearly exact precision, of the stream of psychosocial cues and consequences within the interactions of members of a small group. Under such conditions, systematic investigation can determine which of the many potential modes of communication and verbal interaction are most effective in advancing group stability and efficiency.

Within the context of both analog and simulation settings, a strong case can be made for a strategy of research to address the challenges related to operational performance and general living conditions associated with long-duration spaceflights. Such a research approach will require the integration of investigative research and organizational management activities in an attempt to establish and pretest effective social systems within small groups under operational conditions that allow controlled experimental analyses. In this way, major applied research questions of direct relevance to enhancing the knowledge base required for spaceflight beyond Earth orbit can be pursued without sacrificing methodological rigor by using the participants of primary interest.

A variety of behavioral health and performance research and development opportunities dealing with performance and general living conditions, support and recovery systems, and screening, selection, and training have been described throughout this chapter. These are summarized in Box 5-6. Some will undoubtedly prove to be highly important in contributing to the success of long-duration space missions.

CONCLUSION AND RECOMMENDATION

Conclusion

Behavioral health and performance effectiveness present major challenges to the success of missions that involve quantum increases in the time and the distance traveled beyond Earth orbit.

- *The available evidence-based spaceflight data are insufficient to make an objective evaluation or projection regarding the behavioral health issues that are likely to arise.*
- *The analysis of complex individual and group habitability interactions that critically influence behavioral health and performance effectiveness in the course of long-duration missions remains to be planned and undertaken.*
- *There is a need for more information about support delivery systems at the interface between ground-based and space-dwelling groups.*
- *In the absence of a valid and reliable analysis of the existing database, it is not possible to determine whether the current procedures will be adequate for the screening and selection of candidates for long-duration missions.*
- *Although the data from natural analog environments including simulation studies may be helpful, there remains a need to accumulate knowledge based on observations from systematic research observations*

BOX 5-6
Behavioral Health and Performance Research and Development Opportunities

Astronaut Performance, General Living Conditions, and Group Interactions

1. Enhancing the evidence base on the organization of general living conditions and performance requirements for small groups of humans in isolated and confined microsocieties over extended time intervals and developing an evidence-based approach to the management of harmonious and productive, small, multinational groups whose members will have to function effectively in isolated, confined, and hazardous environments.
2. Coordinating the development of design engineering and habitability requirements on the one hand and evidence-based behavioral health imperatives on the other.
3. Identifying and analyzing those features of small social systems that foster the effectiveness of groups functioning semiautonomously over extended periods of time.
4. Analyzing potentially disruptive group influences that adversely affect harmonious and productive performance interactions under the isolated, confined, and hazardous conditions that characterize long-duration space missions beyond Earth orbit.

Support and Recovery Systems

5. Developing a technology that will provide an adequate means for assessment of the behavioral health effects of long-duration space missions and that will establish and maintain safe and productive human performance in isolated, confined, and hazardous environments and developing an evidence-based approach to the establishment and maintenance of a system for the delivery of behavioral health support and to the analysis of those internal and external factors that influence the effectiveness of the system.
6. Evaluating and enhancing communication with family, friends, and other ground personnel and onboard recreational activities as means of providing behavioral health support for long-duration missions.
7. Evaluating the validity and reliability of performance-monitoring procedures including the Crew Status and Support Tracker, the Windows Space Flight Cognitive Assessment Tool, and the Space Behavioral Assessment Tool and the extent to which methodologies for intramission performance monitoring are enhanced by online downlink capabilities in studies conducted during long-duration missions.
8. Developing and refining procedures for effective intervention under conditions of potentially disruptive personal interactions both among astronauts and between astronauts and Earth-bound support components and for evaluation of the nature and extent of changes in group interaction patterns.

9. Evaluating the effects of ground-based support system design factors including backup components and personnel changes on group integration and stability as they affect personal coaching and technical support functions.

10. Developing and assessing countermeasure interventions that meet the challenges presented by emergencies and technical assistance requirements under conditions with complexities related to cultural and language differences as well as under conditions that involve crews of mixed sexes.

Screening, Selection, and Training

11. Systematically analyzing and evaluating the extensive existing database on the methods and procedures for screening and selection of astronauts used over the past several decades.

12. Evaluating personality measures in the development of valid and reliable procedures for the screening and selection of astronauts and determining the extent to which intelligence and aptitude measures may predict performance more accurately than the more commonly applied personality measures.

13. Developing and evaluating screening and selection procedures that validly and reliably discriminate effective group interaction skills and competences.

14. Developing and refining training technologies including automated training for the preparation of multinational space-dwelling microsocieties as well as their Earth-bound support groups including those related to the interactions of individuals and small groups in the context of distributed ground-based and space-dwelling performance sites.

Data Collection, Analysis, and Monitoring

15. Incorporating and enhancing relevant behavioral health factors as an effective contribution to a more comprehensive plan for the collection and management of astronaut health care data.

16. Developing and testing valid and reliable individualized monitoring and assessment procedures to enhance intrapersonal self-management.

17. Refining communication-monitoring techniques and countermeasure interventions for interactions within and between ground-based and space-dwelling groups.

18. Developing and refining technological approaches to the assessment of individual and group behavioral integrity as well as the efficacies of countermeasure evaluations during long-duration space missions.

19. Establishing a systematic approach to the collection and analysis of postmission (recovery), debriefing, and longitudinal follow-up astronaut health data, including data on behavioral health and performance components.

20. Planning and undertaking a systematic collation and relational analysis of the existing archives of astronaut evaluation data to develop an evidence-based approach to valid and reliable means of screening and selection of candidates for long-duration missions beyond Earth orbit.

in both natural and simulated extreme terrestrial environments and venues like the International Space Station.

Recommendation

NASA should give priority to increasing the knowledge base of the effects of living conditions and behavioral interactions on the health and performance of individuals and groups involved in long-duration missions beyond Earth orbit. Attention should focus on

- **understanding group interactions in extreme, confined, and isolated microenvironments;**
- **understanding the roles of sex, ethnicity, culture, and other human factors on performance;**
- **understanding potentially disruptive behaviors;**
- **developing means of behavior monitoring and interventions;**
- **developing evidence-based criteria for reliable means of crew selection and training and for the management of harmonious and productive crew interactions; and**
- **training of both space-dwelling and ground-based support groups specifically selected for involvement in operations beyond Earth orbit.**

NOTES

The *Mir* with Earth below, as photographed from the space shuttle *Discovery* during space shuttle mission STS-91 on June 12, 1998, during the space shuttle-*Mir* final fly around before bringing the last group of space shuttle-*Mir* astronauts back to Earth. NASA image.

6
Exploring the Ethics of Space Medicine

No data attributable to an individual will be publicly released without written permission of the subject. This concept encompasses non-disclosure of an individual's name and also requires sufficient pooling of data to preclude determining an individual's identity by combining or cross-referencing data.

Institutional Review Board, Johnson Space Center, 1996

Current ethical standards for clinical research and practice with astronauts were developed in an era of short space missions when repeat missions were the norm and a return to Earth within days was possible. In future missions beyond Earth orbit, however, a diverse group of astronauts will travel to unexplored destinations for prolonged periods of time. Contact with Earth will be delayed, and a rapid return will be impossible. Long-duration missions beyond Earth orbit, space colony habitation, or interplanetary travel will create special circumstances for which ethical standards developed for terrestrial medical care and research may be inadequate for astronauts. These ethical standards may require reevaluation.

ETHICAL ISSUES IN CLINICAL CARE FOR ASTRONAUTS

This chapter analyzes ethical challenges in clinical care and research in space medicine as it is currently practiced and suggests a new ethical framework by which space medicine can address these challenges. In particular, the chapter addresses the issues of confidentiality of astronaut medical information, the underlying conflicts in the astronaut-flight surgeon relationship, the difficulties of research with human participants, and the special difficulties that may arise in a long-duration space mission with an international crew.

Institutional Pressure to Underreport Clinical Signs, Symptoms, and Medical Data

Astronauts are subject to a unique set of pressures to keep medical information private. They fear that abnormal preflight medical test results, disclosure of an illness during a space mission, or even evidence of common responses to space travel such as prolonged space motion sickness may be used to disqualify the astronaut from subsequent missions. During its conversations with physician-astronauts in Woods Hole, Massachusetts (see Appendix A), the committee learned that astronauts feel that "the best you can do is come out even" after a meeting with the flight surgeon. The astronauts report that any medical information disclosed to the flight surgeons may jeopardize an astronaut's chance to return to space, particularly if the medical data or the results of those conversations become known outside the examination or interview room.

Because astronauts find the mission selection process opaque (Flynn and Holland, 2000), an undesired mission assignment or a disqualification may be falsely attributed to the disclosure of medical information. The fear of mission disqualification increases their reluctance to report medical information. Rendering the process more transparent by elucidating the procedures and rationales used for astronaut selection for a mission would eliminate medical disqualification as the potential catchall reason for failure to be selected for a mission and increase the likelihood of full reporting of medical information necessary to establish an evidence base for space medicine.

The culture of the astronaut corps, which values stoicism and a "can-do" attitude, further reinforces the individual's reluctance to report medical information. Within this culture illness is often associated with poor job performance or even mental or physical weakness (Flynn and Holland, 2000). Although spaceflight is arduous and requires rising above physical challenges, an atmosphere that equates signs and symptoms (which may be a normal or a common physiological response to an unusual environment) with poor job performance hampers reporting of clinical information about the experience of space travel and thus may limit the attempt to design rationally based countermeasures. Similarly, any tendency to equate medical symptoms with physical or psychological weakness should be eliminated during the National Aeronautics and Space Administration's (NASA's) transition to long-duration space missions.

Underreporting of medical information will limit attempts to establish norms for physiological and psychological adaptations to long-duration space travel and habitation and to design preventive and therapeutic strategies for potentially mission-threatening medical illnesses during long-duration space missions.

Current Practice Regarding the Confidentiality of Individual Astronaut Medical Data

Current knowledge about the physiological and psychological effects of long-duration space travel is surprisingly modest (SSB and NRC, 1998a). Humans have been in space for more than four decades, yet the amount of available data about human responses to prolonged exposure to microgravity remains limited. One of the stated reasons for the limited amount of data that have been analyzed has been a desire to protect the privacy of individual astronauts (Flynn and Holland, 2000). Astronauts, flight surgeons, and NASA staff told the committee that they interpret the Privacy Act of 1974 (5 U.S.C. § 552a [1974]) to strictly forbid disclosure of *any* astronaut medical data in *any* form to anyone outside the flight surgeon-astronaut pair. Whether this is a correct interpretation of the act is unclear, as there are certainly exceptions in medical confidentiality for Earth-based doctor-patient communications (Box 6-1).

Limited doctor-patient confidentiality also exists in dual-agency settings in which the clinician has a duty both to the organization and to the individual. Examples of dual-agency settings include the military, prisons, schools, and other government agencies (Federal Aviation Administration, Central Intelligence Agency, U.S. Department of Transportation, etc.). In any dual-agency setting there is only limited confidentiality to the members of that organization, the members are very aware of these limits, and they sign a Privacy Act statement in their medical records and again in several specific settings (for a psychiatric evaluation, for instance) that document

BOX 6-1
Examples of Earth-Based Limitations on Doctor-Patient Confidentiality to Decrease Risks to Others

1. Control of communicable diseases, as seen in state laws for mandated reporting of resistant tuberculosis, measles, syphilis, and other infectious diseases.
2. Danger or threatened harm to others expressed by patients to physicians (e.g., "Tarasoff" restrictions or child and elder abuse reporting statutes).
3. Occupational health issues that threaten a safe workplace, especially in novel and particularly dangerous workplaces.
4. Restrictions on aviation personnel including military aviators when any symptom or condition might negatively affect safety of flight.

their awareness of the limitations. An example of a portion of the signed Privacy Act statement used in medical records since the mid-1970s in the military is as follows (from U.S. Department of Defense form 2005 of February 1, 1976):

> *Routine Uses:* The primary use of this information is to provide, plan, and coordinate health care. As prior to enactment of the Privacy Act, other possible uses are to: Aid in preventive health and communicable disease control programs and report medical conditions required by law to federal, state, and local agencies; compile statistical data; conduct research; teach; determine suitability of persons for service or assignments; adjudicate claims and determine benefits; other lawful purposes, including law enforcement and litigation; conduct authorized investigations; evaluate care rendered; determine professional certification and hospital accreditation; provide physical qualifications of patients to agencies of federal, state, or local government upon request in the pursuit of their official duties.

At NASA, medical information that is not part of a defined research protocol has been regarded as confidential, to be known only to the astronaut's flight surgeon and the astronaut. This strict stance on privacy also applies to data collected for the Longitudinal Study of Astronaut Health and to all medical information about current astronauts and astronauts in training (Pool, 2000). Because only a few astronauts participate in any given mission, the concern persists that medical data, even if they are presented anonymously, could be linked to individuals. The possibility that an astronaut could be identified is seen as an inescapable barrier to the collection and interpretation of astronaut health data.

Finding: Because of concerns about astronaut privacy, data and biological specimens that might ensure the health and safety of the astronaut corps for long-duration missions have not been analyzed. If these data and other data to be accumulated in the future are to be used to facilitate medical planning for the unique sets of pressures and extreme environments that astronauts will experience on long-duration space missions, the ethical concerns about astronaut privacy must be appropriately modified.

An unfortunate side effect of the overemphasis on individual astronaut privacy may be that future astronauts will be required to undergo observation and experimental investigations that would have been unnecessary if data from past space missions were available and had been analyzed. Repeating research or data collection, particularly if there is a risk to the participants in doing so, adds to the burdens placed on astronauts and is unjustified on grounds of individual privacy. Furthermore, whenever possible all of the data relevant to long-duration travel should be assembled and examined before the design of new protocols for such travel.

Earth-Based Analogs for Balancing Medical Confidentiality with Public Health

There is a long history of balancing individual rights against the health of the public in the monitoring and control of communicable diseases. Disclosure of limited aspects of individual medical data to public health authorities is required for some diseases such as measles and syphilis; in the case of drug-resistant tuberculosis, disclosure of individual medical data can lead to substantial limitations of privacy. These disclosures are based on the following rationale. If a physician is treating a patient for an illness that could affect in a substantive way either the general population or the local population to which the patient will return, the physician is justified in balancing the public health risks against the private needs of the patient. If the public health risk is particularly pressing, the boundaries of confidentiality may be loosened in a manner proportionate to the public health risk.

In another example, a physician may observe that several patients who are all working in a nearby factory each exhibit similar signs and symptoms. The physician may suspect that the work environment is causing or contributing to the health problems. If so, the physician has a duty to investigate the workplace for the source of the illness, even if doing so may require limited disclosure of information about the workers. Again, the severity of the risk to other workers is the standard by which to judge the loosening of patient confidentiality.

Taking these occupational examples further, imagine that the health and safety system for the first nuclear power plant is in the design phase. Because the hazards of working in such an environment could be neither fully understood nor fully anticipated, it would be justifiable to require workers in the plant to wear radiation sensors, to provide routine biological samples to check for radiation exposure effects, and to agree to report signs and symptoms that could be related to radiation exposure to the designers of the health and safety system. Such data would not be released to the general public and should not be released to the employer under the majority of instances, but the data would be essential to those who are in charge of designing the safest possible workplace. Failure of the health officer to know all such data (so that shielding practices could be improved) on the grounds that the data are private could place other workers in danger. The degree of monitoring and the breadth of data gathering should be proportional to the newness of the work environment; more monitoring, and thus less privacy, would be more appropriate in the first 5 plants than in the 20th plant.

Individual astronauts traveling beyond Earth orbit for the first time are quite similar to those workers in early nuclear power plants; there are known and unknown risks to long-duration space missions, and there is a need to create a safe workplace for astronauts on current and future missions.

Justification for Using the Occupational Health Model to Balance Privacy and Safety

Several unique aspects of space travel beyond Earth orbit serve to constrain the presumption in favor of medical confidentiality.

First, the opportunity to observe the effects of space travel on humans is remarkably rare and thus is a scarce resource that should be used or allocated in the most efficient manner possible. In this instance, the rarity of the opportunity to make clinical observations should be balanced against the likely benefits and harms of a strict policy of confidentiality. Second, astronauts will live and work in particular space environments for longer and longer periods of time. The longer the environment is inhabited, the more it will come to resemble a workplace rather than a temporary spacecraft. Issues of occupational health and safety will arise with increasing frequency. In the occupational health context, the confidentiality of the individual's medical information can be judged to be less important when balanced against the safety of the workplace for all "workers" (i.e., astronauts). Commonly in occupational and environmental medicine practice "there is a natural tension between the clinical health issues relating to the individual patient and broader public health issues, which transcend the traditional doctor-patient relationship" (Cullen and Rosenstock, 1994, p. 1).

Several principles of occupational and environmental health practice will apply to any space medicine occupational health model. Modern systems for the surveillance of disease and injury have three components (Baker, 1988):

- data collection,
- data analysis, and
- capacity for response.

In any occupational setting, the broad purposes of occupational health surveillance are the same (Markowitz, 1992). They are to

- determine the incidence and prevalence of occupational diseases and injuries,
- identify individual cases of occupational disease and injury,
- find and evaluate other persons in the same workplaces who may be at risk, and
- discover new associations between occupational agents and disease.

Third, the opportunity to travel in space is highly desired by those in the astronaut corps. Although long-duration space missions will involve substantial hazards, there are likely to be many who will gladly accept the risk in exchange for the unique opportunity to leave the bounds of Earth's orbit. One may rightly view astronauts as they evidently view themselves: a highly privileged group with a rare and valuable opportunity. Such an opportunity brings with it the responsibility to give up some privacy to ensure the safety of fellow astronauts and those astronauts who will follow. If protection of medical confidentiality for individual astronauts hinders the understanding of the risks of long-duration space travel and thus prevents the development of a reasonable medical care program for long-duration space travel, the emphasis on strict confidentiality has been misplaced.

In summary, although the medical privacy needs of astronauts are in some ways similar to those of individuals on Earth, there are models of medical care, most particularly that of occupational health, that permit the ethical collection of individual data if the intention is to design a safer workplace. The work environment of long-duration space missions will be both hazardous and novel. The development of preventive and curative responses to unpredictable hazards will require a refocusing of the usual boundaries of medical confidentiality. Given the small number of astronauts and the small number of space missions, good-faith attempts to protect confidentiality may be more difficult in space than on Earth.

> *Finding: The possibility of identifying an individual astronaut as the source of particular information is not sufficient rationale to continue the current practice of failing to collect and analyze information that could make space travel safer (or even possible) for current astronauts and for those who will journey forth on missions in the future.*

NASA's Clinical Care Capability Development Project (NASA, 1998b; Fisher, 2000) may be an appropriate framework for an astronaut occupational health model to provide optimal health and safety protection of individual astronauts.

The analogy of occupational health carries with it the precedent for certain compromises of individual privacy that will be required for long-duration space missions. A long-duration space mission is, among other things, a traveling space laboratory where experiments, including self-experimentation, are performed. Astronauts in such a traveling laboratory play many roles: a single individual may be both a scientist and a subject, an explorer and a worker, a doctor and a patient. These complicated and multiple roles require rethinking the usual boundaries of medical confidentiality.

ETHICAL ISSUES IN THE ASTRONAUT-FLIGHT SURGEON RELATIONSHIP

Astronauts and flight surgeons often have divergent opinions regarding the boundaries and norms of the doctor-patient relationship in the context of space missions. This is not surprising, given the fact that on the ground the norms of the doctor-patient relationship are deeply influenced by the context of medical care. There are different ethical demands upon both patients and physicians in specialized settings such as an insurance company office, a military battlefield, or a psychiatrist's office. Each of these settings creates variations in the physician's responsibility to the patient and the patient's obligations to the physician. In the context of space missions, the flight surgeon plays multiple roles, from coach, to clinician, to clinical-researcher, to fellow astronaut. The aspects and boundaries of this relationship need to be explored further, not only for their influence on the creation of a safer space travel environment but also for the unstated assumptions that define the roles of astronauts as patients, physician-astronauts in flight, and flight surgeons on the ground (Rayman, 1997, 1998).

OPPORTUNITIES FOR COLLECTING MEDICAL DATA ABOARD THE INTERNATIONAL SPACE STATION

The International Space Station (ISS) will be the main environment for extended astronaut habitation in space during the coming decade or two. How might data about the health experiences of astronauts aboard the ISS be collected to overcome the difficulties in medical data collection and analyses?

Expand the Collection of Occupational Health Data

One important step is to make collection of individual medical data an expected part of participation in all space missions. Monitoring of physiological and psychological data would be expected in return for the opportunity to engage in a unique and desirable activity, namely, the chance to travel in space aboard the ISS. This sort of routine collection of medical data would be part of good preventive health care on any long-duration mission on the ISS or in future missions beyond Earth orbit. Planned analysis of various physical and psychological factors, as well as routine testing to measure the ongoing medical and psychological stresses of space travel, would be required. Reasonable attempts would be made to protect the privacy of any individual astronaut. In many cases, however, the need to design a safer workplace on the ISS and beyond Earth orbit in the future would take precedence over the individual's medical privacy.

A complete listing of the astronaut medical information that would be routinely collected cannot be specified beforehand by this committee. Information that has nothing to do with the ability to travel in space should not be routinely collected, but the committee believes that it would be up to NASA to come up with a policy for routine data collection. It is essential that this policy be created with extensive input from the astronauts. The central assumption of any mandated medical data collection should be that the safety of the astronaut crew takes precedence over individual astronaut confidentiality, although the details of the data to be collected will necessarily change as more becomes known about the hazards of long-duration space travel.

Change the Process for Review of Clinical Research Protocols in Space Medicine

The committee believes that a modified approach to the ethical review of clinical research protocols in space medicine will be required in the era of long-duration missions beyond Earth orbit. Any procedure for review of clinical research performed with astronauts should take into account the unique organizational, cultural, and ethical aspects of the space program. The process for ethical review of research protocols needs to recognize that astronauts are in a unique situation with regard to participation in clinical research. Their situation is quite unlike that of either the healthy or the sick participants in clinical research in typical settings on Earth (Levine, 1986).

In particular, the research review process must recognize the unique pressure on astronauts to overcome any obstacle to participating in space travel. The usual methods of informed consent (Box 6-2) fail to recognize the intense competition among astronauts to be assigned to a particular mission and the implicit coercive effects of such competition.

Although it is NASA policy that refusal to participate in a research protocol should not influence the crew selection process, as stated in Appendix

BOX 6-2
The Common Rule and Informed Consent

The federal policy for the protection of human research participants, referred to as the "Common Rule" (45 C.F.R., Part 46, Subpart A), establishes the ethical framework for all federally funded research with human participants, requiring that research institutions form institutional review boards and delegate to them the authority to review, stipulate changes in, approve or disapprove, and oversee protection of human participants for all research conducted at that institution.

A prominent feature of the Common Rule is the requirement for informed consent. Informed consent is agreement (consent) by the potential participant to participate in a research activity. This agreement is based on an explanation by the principal investigator (the individual who will be conducting the proposed research activity with human participants) in enough detail and in appropriate language to ensure that the potential participant fully understands what he or she is consenting to and that the consent is based on complete knowledge of the nature and risk of the procedure.

V of the handbook of the Johnson Space Center Institutional Review Board (1996), there is evidence that the competition for crew positions renders this policy almost meaningless in practice. Astronauts are reluctant to reveal medical incidents or complaints and to report symptoms during space missions for fear that any physiological variation, however small, may jeopardize their chances of future space travel. If this is the case, astronauts could be expected to sign clinical research consent forms out of the same reluctance, regardless of the official policy. In addition, the methods of crew selection are not so independent of research consent, as the policy suggests. Procedures outlined in Appendix V of the current handbook of the Johnson Space Center Institutional Review Board (1996) acknowledge that a crewmember may be removed from a flight (space mission) for refusing to participate in a specific research protocol "if such action is in the best interest of the flight and the Government" (p. V-2). In addition, such refusal "may prejudice the crew member's assignment to missions of a similar nature in the future" (p. V-2). It is difficult to see how an astronaut could not interpret this to mean that refusal to consent will have substantial negative consequences for her or his career as an astronaut.

Finally, there is clear precedent for modifying policies protecting human research participants to respond to unique or highly unusual circumstances. Perhaps the best-known example is the 1996 exception in the Food and Drug Administration's final rule on informed consent requirements for studies of emergency medical procedures. The exception establishes narrow limits permitting research to go forward without informed consent because, in the emergency or critical care setting, obtaining informed consent often is not feasible. At the same time, the exemption recognizes that patients who have sudden catastrophic illnesses or injuries and who may be unconscious are a special class of vulnerable patients for whom additional safeguards are required. These new safeguards include consultation with the community where the research will be performed, as well as public disclosure of the study design, the risks involved, and the results (Biros et al., 1995, 1998; Menasche and Levine, 1995; Nightingale, 1996).

Suggested Changes in the Approach to Review of Proposed Clinical Research or Data Collection for Astronauts

One approach to the review of proposed clinical research or data collection for astronauts is to divide astronaut clinical research or data collection protocols into two categories (Box 6-3). The first category would include

BOX 6-3
Potential Categories of Clinical Research or Data Collection Protocols for Astronauts

Category 1

Low-risk, noninvasive, and non-time-intensive clinical research and data collection protocols designed to enhance the safety and performance of the crewmembers and the habitability of the spacecraft.

Category 2

Protocols that examine procedures, methods, devices, and so forth that are not likely to be needed during long-duration space travel but that are of value for the testing of early phases of experimental countermeasures or hypotheses about clinical care on Earth.

low-risk and noninvasive research and data collection protocols that are not time intensive. They are designed solely to enhance the safety and livability of all future space travel missions and for the collection of longitudinal health data for astronauts. In such cases, participation in the protocols should be required of all astronauts.

The second category would include protocols that examine procedures, methods, devices, and so forth that are not likely to be needed during long-duration space travel but that are of value for the testing of early phases of experimental countermeasures or hypotheses about clinical care for the general population on Earth (Box 6-4). In this type of protocol, astronauts should have wide latitude in declining or agreeing to participate in research, and no mission selection pressure should be brought to bear to encourage participation. Astronauts would be considered "healthy volunteers" for such

BOX 6-4
Examples of Category 2 Clinical Research or Data Collection Protocols for Astronauts

1. Protocols without a direct effect on space travel.
2. Protocols that examine treatments intended for commercial use on Earth.
3. Those that are high risk or that are invasive or time intensive, even if they are intended to address issues related to prolonged space travel.

research, and strict precautions should be required to make sure the consent process is free of coercion. Informed consent should be most rigorous for studies that are significantly invasive, that are high risk for the individual astronaut, and that offer little potential benefit to the individual astronaut or invade the astronaut's privacy.

Classification of astronaut clinical research and data collection protocols into these two categories will require assessment of the physical and psychological risks of the procedures in the context of long-duration space travel. The clinical research review process should not rely on randomization or the use of particular methods to discriminate between protocols that require consent and those that may be mandated. Instead, the level of risk to the individual astronaut should be balanced against the importance of the results to the creation of safe, long-duration space travel (Truog et al., 1999). In addition, the review process should take into account the physical and in some cases the personal invasiveness of the planned data collection, as well as the other demands on astronaut time, to limit the degree of risk that astronauts must accept. It is only particular types of protocols—low-risk protocols designed specifically to create a safer space travel environment—to which the pressures related to crew selection can legitimately be applied.

If such a two-category system of astronaut clinical research and data collection protocols is adopted, a new clinical research review process for space medicine will have to be developed with greatly increased and active participation from the astronaut corps. The burden of proof for requiring participation in a category 1 protocol should rest with the clinical investigator and the Life Sciences Directorate of NASA, not with the individual astronaut who will participate. The following conditions must be met before participation can be justifiably mandated:

- Astronauts must have a clear sense that space medicine clinical research results will be relevant to the development of a safer space travel environment.
- Astronauts must believe that the space medicine clinical research results will in fact be incorporated into the design of a safer space travel environment.
- Astronauts must be assured that the clinical research protocol will involve the minimal amount of risk possible to attain the desired information.
- Astronauts must be convinced that space medicine clinical research protocols have been designed to have statistical validity and that procedures

or rules based on the results of the research will be implemented within the unique restrictions of long-duration space travel.

• Clinical research results will be openly shared with the astronauts themselves.

Stringent standards must be applied to any clinical research protocol in which astronauts participate, and the review process must be transparent, accountable, and trustworthy from the astronauts' perspective. Application of a dual-agency occupational health model with a clear understanding of the limitations on doctor-patient confidentiality (see Box 6-1 and the paragraph preceding that box) will support a category system of astronaut clinical research with the presence of full and honest informed consent.

ETHICAL ISSUES AND THE SPECIAL CIRCUMSTANCES OF INTERNATIONAL CREWS

The international nature of the ISS presents additional concerns regarding the ethical boundaries of privacy and research. Although there are ongoing attempts to establish international research regulations (e.g., the guidelines of the Council for International Organizations of Medical Sciences [CIOMS, 1993]), little precedent exists for international agreement about clinical and research ethics during long-duration missions in space. The current different approaches of the Russian and U.S. medical programs toward astronaut-patients both on the ground and in space (Williams, 2000) demonstrate that the potential for conflict is strong. One must assume that there will be different approaches to prevention, treatment, and participation in research among nations and that the coordination and organization of data collection will continue to be a difficult task (Levine, 1991).

The committee strongly supports the initiation and continuation of discussions among national space agencies and ISS partners including key stakeholders—that is, astronauts and their families, policy makers, and ethicists—of the ethical principles for clinical care and participation in space medicine clinical research. It would be unacceptable to assume that all members of an international crew will share the same views about research participation or the release of medical information. Nor would it be acceptable to limit the collection of data from particular ISS crewmembers simply because of a failure to discuss the ethical boundaries of data collection before that crewmember's launch into space.

CONCLUSION AND RECOMMENDATION

Conclusion

The ultimate reason for the collection and analysis of astronaut health-related data is to ensure the health and safety of the astronauts.

- *Emphasis on the confidentiality of astronaut clinical data has resulted in lost opportunities to understand human physiological adaptations to space, and concern for the protection of privacy and over the implications regarding disclosure and use of clinical data may have led to the underreporting of relevant information.*
- *Reevaluation of the application of the Privacy Act and statutory privacy provisions may be necessary to enable appropriate access to necessary data while protecting the privacy of the individual astronaut.*
- *The unique environment of deep space, combined with the social and institutional contexts of health care research with astronauts, requires that astronauts be considered a unique population of research participants.*
- *A limited international consensus exists on the appropriate principles and procedures for the collection and analysis of astronaut medical data. The potential for conflict among the national space agencies and International Space Station partners is high.*

Recommendation

NASA should develop and use an occupational health model for the collection and analysis of astronaut health data, giving priority to the creation and maintenance of a safe work environment.

- **NASA should develop new rules for human research participant protection that address mission selection, the limited opportunities for research on human health in microgravity, and the unique risks and benefits of travel beyond Earth orbit.**
- **A new interpretation or middle ground in the application of the Common Rule (45 C.F.R., Part 46, Subpart A) to research with astronauts is needed to ensure the development of a safe working environment for long-duration space travel.**
- **NASA should continue to pursue consensus among national space agencies and International Space Station partners on principles and procedures for the collection and analysis of astronaut medical data.**

The partially assembled International Space Station *Alpha* backdropped against clouds and snow-covered mountains photographed from the space shuttle *Discovery* during STS-102 on March 1, 2001. NASA image.

Artist's rendition of the fully assembled International Space Station *Alpha*. The pressurized volume of the station following a 5-year, 45-mission sequence of assembly will be roughly equivalent to the space inside two Boeing 747 jet airplanes. NASA image.

7
Planning an Infrastructure for Astronaut Health Care

When we are ready to send humans deep into space, astronaut health and safety will be our top priority.

Daniel S. Goldin, NASA Administrator
Speaking at the U.S. National Academy of Sciences
October 25, 1999

MISSION AND GOALS

Space is the most extreme environment that humans have ever entered. A prolonged presence in microgravity results in a number of adaptations that may not be completely reversible. Space crews are isolated and spacecraft provide limited room in which to live and work. Future exploration of the solar system may require crews to land on other planets (such as Mars) or to establish colonies on other bodies in space (such as the Earth's Moon). Currently, radiation protection in deep space is an unsolved issue, and the threat of dysbarism is ever present, especially during extravehicular activities. Traditionally, the National Aeronautics and Space Administration (NASA) has depended on a preventive approach to astronaut health, reflected in strict astronaut selection standards and close monitoring of astronaut health status. This has proved adequate for short-duration space missions, and NASA crews have been essentially free of major illness during the missions. To date, few urgent medical problems have occurred, and the spacecraft has always been within a day or several days of return in case of a serious emergency. This may not be the case during long-duration missions.

With long-duration space missions the number and variety of health conditions facing astronauts who may require medical intervention before, during, and after space travel will continue to expand. There are medical conditions that have the potential to seriously impair the ability of astronauts to function in the spacecraft, during egress, while working outside the spacecraft, and upon the return to Earth. With increasing mission lengths, the chance of an acute medical emergency such as appendicitis, pneumothorax, or acute mental impairment increases, as does the necessity for medical intervention within the spacecraft. Relevant questions include the following:

- What are the physical and behavioral attributes best suited for long-duration space travel?
- How do the series of preventive measures necessary for a long trip differ from those designed for shorter space travel?
- How should methodologies and technologies for care evolve?
- Are there health care issues especially pertinent to landing or colonization that are different from those pertinent to extended travel in a spacecraft?

To support extended space travel, NASA's medical care system must, at a minimum, maintain the health of each crew member so that she or he can (1) function as a productive member of the crew, (2) retain acceptable health during and upon completion of the spaceflight, (3) egress from the spacecraft, and (4) maintain orthostatic tolerance during deorbiting and landing (briefing, Johnson Space Center, February 2000). Although these goals are consistent with those of space travel of any duration, how they are attained differs. For example, for missions of the space shuttle, what is reasonably required is modest ambulatory care, first aid, and rudimentary "life support." On the International Space Station (ISS), additional components, including more advanced life support, stabilization, and transport, are necessary. Beyond Earth orbit, there is the further necessity of definitive treatment, rehabilitation, and chronic care because of the inability of return to Earth for many months. The methods necessary to achieve these goals are not yet developed. In addition, the full array of health conditions that must be prepared for, although under intense study by NASA and described in part in Chapters 3 and 4 of this report, have not been fully catalogued, nor have those that have been identified been addressed adequately.

Why does NASA need a comprehensive health care system for space travel? There are two reasons: deep space is a unique environment with special hazards for humans, and at this point, humans are thought to be necessary for an exploratory mission. The environment is unique in several ways. The most salient one is that humans have not been beyond Earth orbit for prolonged periods, and as mentioned in other sections of this report, not enough is yet known about the effects of prolonged exposure to microgravity on humans to send them there safely. For this reason the Institute of Medicine Committee on Creating a Vision for Space Medicine During Travel Beyond Earth Orbit has strongly recommended the use of the ISS as a platform for clinical research on the effects of microgravity and for validation of countermeasures. In addition, technological problems, such as radiation protection, remain unsolved, making long-duration space travel probably unacceptably dangerous. Finally, although all previous voyages of discovery on this planet have had, to greater or lesser degrees, the availability, outside the craft, of food, water, and air and a means of waste disposal, the impossibility of a reasonably speedy return means that all that is necessary for prolonged maintenance of life must be contained within the spacecraft.

Although there is debate within the scientific community of the necessity of sending humans on long-duration exploratory space missions, the committee assumes for the purposes of this report that humans will be sent. Given that assumption, there is the obvious necessity to keep them functioning productively. In addition, there is the ethical imperative that they return to Earth in acceptable health. The committee strongly believes that the success of the mission and the successful return of healthy individuals demand, on the basis of risk analysis and management of the evidence base, the highest reasonably attainable standard of health care with the resources available before, during, and after such a mission.

ORGANIZATIONAL COMPONENTS

Providing the highest reasonably attainable standard of health care for astronauts during long-duration missions beyond Earth orbit will require an effective organizational framework that integrates health care for astronauts with an effective health care research strategy.

The committee reviewed two documents that provide a portion of the conceptual framework for this section. The two documents, *Resources for the Optimal Care of the Injured Patient* (ACS, 1999) and *Model Trauma Care System Plan* (DHHS, 1992), address Earth-based care for a specific category of patient: patients with acute injuries. These documents are relevant because they describe acute care for the most urgent medical needs that may occur during space travel and because they describe the framework or infrastructure that an adequate health care system must have to meet those needs. The documents also represent examples of what may be learned from fields outside those to which NASA has normally turned for advice and, in the case of the document from the American College of Surgeons, emphasize the importance of periodic updating of standards and procedures on the basis of new knowledge and theory.

Leadership

The committee's perspective on leadership is based on the training and experience of its members. That background emphasizes two values relevant to the issue of leadership: that someone is in charge and that continuity is a key component of health care. "In charge" requires both accountability for and authority over the funding needed to make all pertinent decisions with respect to the health of the astronaut corps. "Continuity" means that the health care system provides for the continuity, over time and space, of the health care of astronauts.

It is the intent of the committee to address the principles that it believes are necessary for an effective system for astronaut health care. The committee intends no judgment, positive or otherwise, of the current structure or of the individuals currently holding positions within NASA. Indeed, the NASA organizational structure dealing with astronaut health underwent substantial changes during the time of the committee's study. These include a Johnson Space Center (JSC) proposal for a Bioastronautics Institute to coordinate research and health care and the designation of a NASA chief medical and health officer to provide health and medical policy oversight. Although the committee renders no opinion on the validity of these specific steps, it commends NASA for recognizing the need for change and believes that these steps could be appropriate in meeting the concerns that the committee developed in the course of its discussions. The committee is concerned, however, that the present organizational structure has not yet evolved to the level necessary to ensure astronaut health and safety during travel in deep space.

Critical Elements of the Organizational Framework

Developing a framework within which a health care system for astronauts can evolve is best accomplished through the designation of a single organizational component. Whether this component is internal or external to NASA, it should be headed by an individual with the appropriate training, experience, and authority (1) to develop the system along with standards for performance; (2) to coordinate all related external and internal resources, including basic, translational, and clinical biomedical and behavioral research; and (3) to administer the component's policies and procedures. The committee is concerned that fragmentation of the necessary elements of the framework will work to the detriment of the health of the astronauts. Important to the committee in this regard are the elements of *coordination* with other organizational units within NASA, *integration* of astronaut health with other components of the space mission, and *authority* to determine what is in the best interests of astronaut health.

Coordination

The committee learned of numerous examples of a lack of coordination among different organizational components of NASA. One pertinent example, also commented on by the Space Studies Board (SSB and NRC, 1998a), is the lack of coordination among the Ames Research Center, JSC, and National Space Biomedical Research Institute (NSBRI) for countermeasure development, the process that NASA views as being critical in providing answers to unsolved clinical questions. It is understandable that in any large and complex organization different components may work on similar issues. It is also notable that JSC and NSBRI have begun to coordinate countermeasure development. Nevertheless, for many reasons, including that of the difficulty of making general and prospective valid conclusions from a small number of observations or with a small amount of data (the problem of studies with small numbers of participants, that is, the "small *n*" problem mentioned in Chapter 2), it behooves NASA to coordinate all activities related to astronaut health.

Integration

The work of NASA originated in engineering, and its most stunning successes have been technological. Indeed, the committee heard in discussions with astronauts that even the most biologically oriented of astronauts, the physician-astronauts, identify themselves first as astronauts and then as physicians. In discussions with NASA officials, staff, and astronauts, however, the committee learned of several instances in which the human component of space missions was given, at best, short shrift. For example, the committee believes that the habitability of the spacecraft becomes a critical issue on missions that could take as long as 3 years. The committee believes that improvements in this area, as well as many others, would benefit from a greater degree of integration among the various components of NASA. As the committee has noted throughout this report, biology and engineering are moving closer together, in theory and in practice, as nanotechnology benefits damaged biological systems and as the biological concept of self-repair enters engineering. This transition represents an opportunity of historic proportion, but its benefits require organizational integration.

Authority

The risks to the health and the lives of astronauts on long-duration missions are substantial. That factor alone should require the establishment of an organizational component with a named individual accountable for astronaut health. The organizational component should have significant and substantial input from the astronauts. The organizational component would be even more effective and would be trusted by astronauts even more if the astronauts themselves had some input into the selection of the individual responsible for their health and some ongoing input into that individual's performance review. Astronaut input would be in an advisory and consultative capacity, not in a controlling or managing capacity, and would give real credibility to the stated commitment to make astronaut health and safety the highest priority of the mission. Furthermore, the individual accountable for astronaut health must have authority sufficient to ensure that all appropriate steps are taken to maintain the health of astronauts.

The committee believes that the element of authority is of sufficient importance to require that the individual have both operational authority and budgetary authority, with checks and balances provided in part by internal NASA mechanisms and supplemented by an external advisory group modeled on advisory groups of the National Institutes of Health (NIH) and other federal external advisory groups, for example, those at NASA, the Office of Naval Research, the U.S. Department of Defense (DOD), and the Environmental Protection Agency. The qualifications of the individual should include a knowledge base in space medicine, experience with the administration of a health care system or a major component of a health care system, and a clear understanding of the important role of research. Given the relatively limited opportunities to do clinical research in space, efficient research prioritization and administration are critical.

Alternative Organizational Frameworks

The committee considered alternative frameworks within which an organizational component responsible for all aspects of astronaut health might evolve. The framework could be external (for example, it could be a component of a larger external health care system, established through a contractual arrangement, or part of a public-private group such as NSBRI), or it could be internal to NASA and could be located at JSC or NASA headquarters. Each type of organizational framework has its own advantages and disadvantages.

External Frameworks

Astronaut health care as a component of a larger health care system NASA could partner with an existing federal health care system with specialized knowledge applicable to NASA's needs in aerospace medicine, such as the one within DOD, or it could partner with the Federal Aviation Administration (FAA). The National Science Foundation (NSF) has in the past used resources of the U.S. Navy to provide health care during NSF long-duration overwintering missions in the Antarctic. The advantages of this type of arrangement would include the use of an established, comprehensive, and evolving working system. The disadvantages are those associated with the use of any external system, for example, a lack of or a decreased familiarity with and specificity for the needed task, a lack of flexibility to evolve with space medicine, a lack of confidentiality, limited degrees of budgetary authority and policy-making ability, and a lack of familiarity with the international space medicine community. The greatest disadvantage of an external system is its diminished capability to coordinate with the health care research strategy of NASA to produce clinically relevant information so that NASA can provide the highest reasonably attainable standard of health care for astronauts during long-duration missions beyond Earth orbit.

Astronaut health care established through a contractual arrangement NSF is providing health care in Antarctica through a contractual arrangement. Along the same lines, NASA is using Wyle Laboratories to manage its Countermeasure Development and Validation Project and could expand this arrangement to include management of a comprehensive health care system for astronauts. Although the existing arrangement with Wyle Laboratories is an advantage in terms of the contractor's familiarity with the task, Wyle Laboratories' specialized focus on countermeasures is a potential conflict, and no NASA contractor-based health care system that could be built upon is in place. Contracting with an established nonfederal health care system or organization to provide comprehensive health care for astronauts would have a mixture of the advantages and disadvantages described above for an established federal health care system, particularly the diminished capability for integration of clinical information and clinical research into the health care research strategy of NASA.

Astronaut health care as part of a public-private group NSBRI is responsible for research in countermeasure development. Its task could be expanded to include the management of a comprehensive health care system for astronauts. Its university members provide expertise in a wide variety of areas, although they are selected to provide expertise in basic and translational clinical research. The resources of an established health care system, however, are not present and would have to be developed; and the potential disadvantages described above for partnering with an established federal health care system external to NASA would also be present.

Internal NASA Frameworks

The existing components within NASA that provide health care for astronauts could continue to evolve into a comprehensive health care system for astronauts. The physical facilities and basic organizational structure proposed for the Bioastronautics Institute, described to the committee during its visit to JSC, indicate that the Bioastronautics Institute would be an attractive alternative as the organizational component that could provide comprehensive health care for astronauts. NASA should continue to study this alternative. It would represent an evolution of existing programs, would be close to the astronauts, and would likely be able to gain astronauts' confidence more easily than an external system would. It would be more amenable to providing flexibility, interpersonal communication, and the degree of confidentiality needed. It would also have more budgetary, operational, and policy-making authority and would have a greater ability to coordinate with the external research and international communities than an external organization would.

NASA has several frameworks to consider as it continues to build a comprehensive health care system for astronauts. Whatever framework NASA decides upon, it must lead to reduced fragmentation and increased communication at all levels and must provide the highest level of health care for astronauts.

Organizational Structure to Ensure Astronaut Health and Safety

The committee's logic in its assessment of the organizational structure needed to ensure astronaut health and safety is as follows: a voyage of exploration into deep space that would last for up to 3 years is of a different order of magnitude and is a different kind of mission than has ever been attempted. Without effective countermeasures developed from basic knowledge of biological and clinical processes, humans would likely not survive such a mission. For the past 40 years and to the present, NASA's organizational structure has missed critical opportunities to collect and analyze relevant human clinical data, data that if collected and analyzed would likely have placed NASA closer to solving the necessary clinical and engineering issues.

The committee believes and reiterates here that to ensure a successful long-duration mission into deep space, all that can be done to collect and analyze relevant clinical and basic biomedical data must be done. Thus, the committee has recommended a two-pronged strategy of (1) development of a comprehensive health care system to collect clinical epidemiological data and (2) development of a strategic biomedical and clinical research plan to collect and analyze data on health risks and their amelioration. Integration and coordination of these two components are, in the committee's view, absolutely necessary. Such integration and coordination are best overseen by a single qualified individual with sufficient authority.

After considering the evidence and testimony presented to it during the meetings conducted for preparation of this report, the committee strongly believes that a major revision of NASA's organizational structure is needed to provide for the optimum health and safety of astronauts for future long-duration missions beyond Earth orbit. The committee believes that NASA needs a single organizational component whose head has overall responsibility for the health and safety of astronauts. Operational or line responsibility for astronaut health care and clinical research would likely rest with two different individuals, who would report to a single individual. That individual's responsibility would be to coordinate, oversee, and set policy for all astronaut health care and clinical research, similar to the lines of authority and organization at the U.S. Department of Veterans Affairs and at NIH. In reaching this decision, the committee considered several alternative organizational possibilities, as described above. The committee believes that a single organizational component within NASA has the best potential to achieve the evolving framework needed to coordinate with other organizational units within NASA and integrate all aspects of astronaut health and safety with other components of the space mission. The unit would also possess the authority to make decisions in the best interests of astronauts and their health. The committee is mindful of the fact that it is advancing a major structural change that may involve shifts of power and that will involve readjustment of significant resources. The change will, in the considered judgment of the committee, allow NASA and the international space community to build a better-integrated comprehensive system of health care to provide for the best possible health and safety of all astronauts who travel beyond Earth orbit.

CONCLUSION AND RECOMMENDATION

Conclusion

The challenges to humans who venture beyond Earth orbit are complex because of both the unique environment that deep space represents and the unsolved engineering and human health problems related to long-duration missions in deep space. The committee believes that the current organizational structure of NASA may not be appropriate to successfully meet the challenge of ensuring the health and safety of humans on long-duration missions beyond Earth orbit.

- *Astronaut health and performance will be central to the success of long-duration space missions, but the responsibility for astronaut health and performance is buried deep within NASA.*
- *Within NASA the focus on health care research and astronaut health care is not sufficient, nor does NASA sufficiently coordinate and integrate the research activities needed to support successful long-duration missions beyond Earth orbit.*

Recommendation

NASA should establish an organizational component headed by an official who has authority over and accountability for all aspects of

astronaut health, including appropriate policy-making, operational, and budgetary authority. The organizational component should be located at an appropriate place and level in the NASA organizational structure so that it can exercise the necessary authority and responsibility. The official who heads the organizational unit should be assisted by officials who are separately responsible for clinical care and health care research. The proposed organizational component should

- **have authority over basic, translational, and clinical biomedical and behavioral health research;**
- **foster coordination between NASA and the external research community; and**
- **be overseen by an external advisory group, modeled on advisory groups of the National Institutes of Health and other federal external advisory groups, to provide program review, strategic planning, and leverage to assist NASA in meeting its goals for astronaut health.**

Systems Development

To develop a successful functional health care system for astronauts, two components are necessary: systems planning and system operations (Box 7-1). These processes afford an opportunity to systematically develop the elements of a comprehensive system and to allow consensus to build around the program.

BOX 7-1
Infrastructure Elements for Developing a Comprehensive Health Care System for Astronauts

Planning

1. Assessment of needs and resources.
2. Process for development and implementation of a health care plan.
3. Process for standards development.
4. Implementation plan, with an approval process and timetable.
5. Description of how both interested parties and external advisers will participate.

Operations

Clear description of the process for establishing, implementing, and updating policies, procedures, and protocols.

Planning

NASA already has in place a process, the Critical Path Roadmap project, in which astronaut health care issues are identified and mechanisms for addressing those issues are developed. A similar process could profitably be applied to the development of an infrastructure for a health care system for astronauts. Such a process could be used to identify needs and evaluate resources.

A process for development and implementation of a health care system for astronauts that covers the continuum from premission, mission and postmission and that provides for systematic review should be established. A systematic review process is especially important since although the infrastructure for a comprehensive health care system for astronauts may be put in place in the near term, the implementation of portions of a comprehensive health care system for astronauts during long-duration missions is decades away.

Standards of care must be developed. Although a process is under way within NASA, there is not yet a clear statement of what the standard of health care for astronauts on long-duration missions should be or what practical limitations to that standard may exist. Perhaps most vexing is the probable international nature of any long-duration mission and the different ways in which different cultures view risk. Thus, the goals should be defined with all appropriate parties involved. An implementation plan, with an approval process and a timetable, should be developed.

Most important, in the committee's view, is a description of how all interested parties and external advisers will participate in the planning process. The reason for this is based largely on the committee's discussions with NASA staff and on anecdotal evidence that NASA as an organization is perhaps more insular than it should be. The committee believes that the interaction of humans and machines will be even more critical during long-duration space missions than on short-term spaceflights.

Operations

There should be a clear description of the process for establishing, implementing, and updating policies, procedures, and protocols. This should include descriptions of how the different operational units will relate to each other and how their activities will be integrated. In addition, the mechanisms for cooperation with other components of government should be described. NASA should continue the series of cooperative agreements with NIH, DOD, and other agencies that it has already instituted.

Policy and Legislation

Given the committee's perspective that long-duration missions beyond Earth orbit present unique issues, the time appears to be appropriate to review the legislative authority and legislative interpretation of NASA's programs to ascertain whether some modification is advisable. Three areas may be especially appropriate.

Health Data

A principal recommendation of the committee concerns the acquisition and treatment of astronaut health data and the application of occupational health principles to those data. The gathering of health data and their analysis are crucial to the success of being able to provide appropriate health care. NASA staff have cited two federal statutes, the Privacy Act and the Freedom of Information Act, and principles of medical confidentiality as constraints on the collection and full analysis of astronaut clinical data. Either the acts themselves or overly conservative interpretation of them has limited the acquisition of data necessary for the protection of astronauts and future space crews. If it is the former, consideration should be given to modification of the statutes in a way in which "need to know" is the guiding principle. If it is the latter, NASA should reevaluate its interpretation of the acts. NASA should also consider using the reporting requirements for occupational health data.

External Advice

Elsewhere in this report the committee expresses the concern that NASA has not cast a wide enough net in seeking external advice. Although a number of advisory committees exist throughout NASA, including the NASA Advisory Council and the Life and Microgravity Sciences and Applications Advisory Committee, advice related directly to medical issues appears to come from the Medical Policy Board. The Medical Policy Board is composed solely of government employees, mostly physicians. The committee suggests that NASA consider whether modifications to this policy structure is advisable by including members who are not federal employees to provide broader expertise and advice in ensuring protection of astronaut health and ensuring astronaut safety.

Clinical Research Plan

One of the committee's principal recommendations is for NASA to develop a strategic clinical research plan. Presently central to NASA's biomedical research program, including its clinical research program, is NSBRI. NSBRI, chartered as a nonprofit corporation under Texas law, includes a consortium of academic institutions; it also has relationships with industry. The committee believes that NSBRI has the potential to foster fundamental research into the critical clinical questions facing astronauts on long-duration missions beyond Earth orbit. Although the committee makes no recommendation concerning the detailed research structure of NASA, NASA should consider whether the current structure of NSBRI and its legislative authority are appropriately effective in achieving NASA's research goals.

Funding

The total fiscal year 2000 funding for NASA-supported programs in biomedical research and countermeasures was approximately $57 million. This figure does not include support for NSBRI from other sources, nor does it include research sponsored by other agencies of the federal government. Measured against the overall NASA budget of nearly $14 billion, in which technological systems dominate, the amount is small. Given the committee's conclusion that unless critical clinical problems are solved humans cannot safely "fly" on long-duration missions beyond Earth orbit, NASA should reconsider its research funding priorities.

Internal Relationships

The role of the human component of space missions needs to be upgraded. By this, the committee intends three points: (1) engineering and biological perspectives need to be brought into better balance, (2) sufficient priority needs to be given to learning what is necessary to send humans safely on long-duration missions, and (3) the success or failure of long-duration exploration missions beyond Earth orbit is dependent upon humans. As good as technology is (and it is quite good), successful missions continue to depend on human expertise and ingenuity in solving problems. Examples include the successful return of *Apollo 13*, the repair of the Hubble Space Telescope, and most recently, successful deployment of the solar panels on the ISS.

The committee suggests that NASA consider whether its current structure and priorities adequately reflect the importance of the human component in space. In addition, it should consider whether organizational coordination, integration, and authority meet the criteria suggested in the section on leadership above.

External Relationships

It is the opinion of the committee that the importance of external relationships in accumulating necessary data, communicating necessary information, dealing with international partners, and obtaining advice is worthy of comment.

Accumulating Necessary Data

Expanding capabilities, freeing resources to do what only NASA can do easily, building more inclusive networks, staying abreast of the newest innovations and relevant trends, encouraging others including the next generation to participate in development of the field, and open and strong public communications are essential to the success of any large endeavor like human exploration of deep space. Until relatively recently, NASA had not used other agencies of the federal government for the accumulation and development of clinically important information. The committee believes that cooperative agreements with NIH should continue and should be expanded. For example, there is intense interest in many NIH institutes to increase knowledge of osteoporosis and its causes. At present and in the past there has been considerable interaction between NASA, the National Institute on Aging, and the National Institute of Arthritis and Musculoskeletal and Skin Diseases. These interactions should be encouraged and enhanced to address the important issue of bone loss during space missions and to help provide information about osteoporosis in general.

The research resources of NIH, both in terms of funding and in terms of infrastructure for biomedical research, are far greater than those of NASA and could appropriately be used to support basic and clinical research that meets the priorities of both institutions. Not only can NIH-sponsored research help address clinical issues pertinent to long-duration space missions, but NASA-sponsored research in microgravity also can benefit NIH's mission in fundamental biomedical research. Another agency, in addition to DOD, that has been involved in aviation-related data collection is FAA. FAA, whose aviation medicine component regulates and approves medical certification for civilian aviators, might have clinically important information that NASA could use.

Although NASA should be very proud of its accomplishments in space medicine, more formal collaborations with NIH and perhaps other research agencies within the U.S. Public Health Service and the federal government would be well worth considering. Some of these collaborations might be at a simple administrative level (greater use of joint study sections to review grants and contracts), and some might be accomplished by greater involvement with NIH intramural programs through formal interagency agreements. There is also perhaps a need to focus on greater collaborations with the private sector in matters relating to genotyping, gene expression, and proteomic profiling in the context of space medicine.

There is a need for information on the actions of pharmaceuticals in microgravity. As noted by the National Research Council's Space Studies Board and in information provided to the committee, there have been anecdotal reports of altered drug efficacy on short-duration space missions. Although the Space Studies Board did not address clinical pharmacology in detail, it did recommend carefully designed clinical research. No data exist on whether drug action in microgravity is similar to that on Earth or which drugs may be most appropriate for inclusion in a formulary. Because of the increased importance of curative pharmaceuticals on a long-duration mission, cooperative arrangements with the pharmaceutical industry would appear to be appropriate and cooperatively designed experiments would appear to be mutually beneficial. The initial focus of cooperative research in this area would not be efficacy but would be whether the actions of existing efficacious drugs are altered in microgravity.

Development of an onboard formulary is a different, but no less important, matter. It will be impossible to carry every possible pharmaceutical agent on a long-duration mission. Thus, it will be necessary to choose what to take and, possibly in the future, to determine what will be able to be synthesized on a mission. Many organizations both within and especially outside the federal government have addressed formularies for their own patient populations and presumably have processes for formulary development. The U.S. Department of Veterans Affairs, the DOD, and many university-related managed health care plans might offer helpful advice.

Information development applies not only to scientific or clinical knowledge but also to knowledge of processes and models. NASA has begun to develop clinical protocols, which are of tremendous value in clinical practice. Because there may be limited clinical experience among the astronauts within the spacecraft and because of the significant delay in radio or even more advanced means of transmission during missions beyond Earth orbit, it is anticipated that clinical protocols (practice guidelines) will be the basis for medical decision making. Significant advances in guidelines development have occurred in the past two decades. The Agency for Healthcare Research and Quality (AHRQ) within the U.S. Department of Health and Human Services has been a leader in this area; cooperation between NASA and AHRQ could enhance NASA's work in this area. In addition, DOD has substantial practical experience with clinical protocols. The committee heard specifically of the utility of clinical protocols administered by corpsmen in submarines. Further collaboration between NASA and DOD would similarly be useful, as would collaboration with the U.S. Department of Veterans Affairs with its experience with clinical guidelines, clinical protocols, informatics, and performance measures.

The health care systems for individuals in extreme isolated terrestrial environments—Antarctica, submarines, and the deep sea where divers work (see Appendix A)—share common structural characteristics. The committee believes that NASA could continue to learn from these analogs in developing its own health care system. In addition, the branches of the armed services have a long history of working in extreme situations and in developing model systems (IOM, 1999a). For example, most of what is known today about trauma surgery was developed on the battlefield. Furthermore, the military model of the organization and delivery of medical care may provide further insights into NASA's nascent organization of a model for health care during long-duration space missions.

Communicating Necessary Information

Although exploration and discovery hold special places in the human psyche, space holds a fascination. Symbols of that fascination abound, from the National Air and Space Museum, to the NASA-dedicated cable television channel, to hit films such as *Apollo 13*, to programs about space on The Discovery Channel. Nor was the committee immune to that fascination. NASA has done an excellent job in building on this fascination, from the first lunar landing to the construction of the ISS.

NASA should pay increased attention to two areas of public communication, however. These are the communication of the benefits of research in space and the communication of risk. Both research support and research cooperation are often dependent on the perception of benefit. The committee heard from NASA staff that support for biomedical research was lacking from the scientific and political communities. The committee believes that cooperative agreements, coupled with a clear statement of the benefits of research in space medicine, would enable NASA to gather additional support for the necessary research.

Much more important in the view of the committee is the need to communicate risk. Spaceflight is inherently risky. Long-duration space travel, especially that beyond Earth orbit, entails known as well as unknown risks. The committee believes that there is a profound professional and ethical responsibility to evaluate honestly the risk to human life that will be incurred as a result of extended space travel. Although the risks must be evaluated in the context of the benefit to humans, they must be stated on an individual human level in terms that can be plainly understood. NASA states that the issue of risk is a high priority. The determination of what risks to humans exist and what countermeasures should be developed is being addressed initially through NASA's Critical Path Roadmap project. Nevertheless, the present committee report touches on other information that must be developed and other mechanisms that should be used before the goal of optimally achievable safety to humans on long-duration missions is approached. Of equal significance is how, in a truly informed and deliberative way, an individual astronaut may come to a personal decision to accept the risk of a maiden voyage to, for example, Mars. This issue is further discussed in Chapter 6.

From the perspective of society, the committee believes that risks and benefits must be addressed and communicated explicitly. The public should be made aware of the risks to astronauts, as well as the benefits, of space missions. Five centuries ago, a voyage west from Europe was, to some, a voyage off the face of a flat Earth, a voyage necessarily doomed to tragedy. A century ago, Antarctic explorers faced clear odds against survival. Today, for every ten people who successfully summit 8,000-meter peaks, one or more dies (The New York Times, 2001). Every day individuals voluntarily take risks of a degree that they would not take if another party required them to take those risks. Yet, many in the United States assume that there are activities that should be risk-free. The public must be prepared for the possibility that all countermeasures may tragically fail and that a crew may not return from a prolonged space mission. The public must also be prepared for the possibility that on the astronauts' return to Earth, normal functioning in some areas may not return for a long time and in some cases may possibly never return. Although NASA has identified risk as an issue that it needs to address, the committee concludes that a greater emphasis must be placed on communicating health risks to astronauts and to informing society not only of the benefits but also of the risks of space travel.

Dealing with International Partners

Given the uniqueness and complexity of long-duration missions beyond Earth orbit, it is anticipated that there will be a need for increased international cooperation to achieve success. What is learned on the ISS will be applicable to such long-duration missions. The committee's recommendation that the ISS be a platform for clinical research stresses the importance of the following recommendation of the Space Studies Board (SSB and NRC, 2000, p. 78): "Mechanisms are needed to ensure that protocols and facilities for pre- and postflight monitoring and testing are consistent across national boundaries. There have to be common criteria for evaluation and utilization of countermeasures and international cooperation in their development." The committee would add that medical care, as practiced on the ISS and on long-duration missions beyond Earth orbit, must also be consistent for individuals of all nationalities.

Obtaining Advice

On the basis of the committee's recommendation that a comprehensive system for astronaut health be established, it recommends that a NASA external advisory board, committee, group, or panel whose sole concern is astronaut health be established. Because of the small *n* problem and the limited amount of research that can be carried out on the ISS and during space shuttle flights, it will likely never be possible to develop a knowledge base of information important to astronaut health to the level of validity and reliability to which researchers may be accustomed in other circumstances. The small *n* problem, and approaches to solving it, are discussed in the IOM report, *Small Clinical Trails: Issues and Challenges* (IOM, 2001e). An external advisory group on astronaut health whose role is to act as ombudsperson for astronauts and that has the expertise to understand not only the complex issues of space medicine but also the importance of continuing research would serve as a check and balance to the inappropriate use of astronauts during missions. It would also support when missions must go forward even in the absence of perfect information. Such an advisory group, external to NASA, could also ensure that

- the health and safety of the astronauts are priorities;
- operations are rigorous in terms of health care system design and ongoing modification;
- research is coordinated and appropriate;
- data collection, retrieval, and analysis are seamless and data are linked to policy and practice;
- methodologies of continuous quality improvement are used; and
- regular reports on the health care program itself are provided to the full NASA community.

A separate reason for an external advisory group on astronaut health is the need to stay abreast of new and emerging scientific knowledge. The two decades before an interplanetary mission will likely occur will see not only massive amounts of new and relevant basic and clinical information but also the coming and going of concepts and practices of health care delivery and technologies that support that care. Clinical practice guidelines cannot be static but must be updated on the basis of new knowledge, therapies, and technologies. A current example is the Human Genome Project and the effect that it will have on prevention, diagnosis, treatment, and rehabilitation. Other, relatively new, and rapidly developing fields of nanotechnology, noninvasive means of diagnosis and therapy, medical informatics, and so forth have the potential to transform the approach to medical care in space as well as on Earth. NASA should have a mechanism for systematically tracking and analyzing not only the research with which it is directly involved but also research occurring in other fields that may be relevant to its missions.

To accomplish this, collaboration with other research agencies within the federal government, all of which have some degree of incentive to track and compile data from research, would be beneficial. An additional approach is the bringing together of "skunkworks," an eclectic group of individuals that brainstorms on particular areas and topics on which they are focused as they emerge and a group that, constantly renewed, would have an ongoing advisory function. The committee suggests that NASA look at the role, structure, and usefulness of external advisory groups at NIH and other federal agencies as a guide.

OPERATIONAL AND CLINICAL COMPONENTS

On the basis of the committee's perspective that deep space represents a unique environment and that the health and safety of astronauts are paramount, the committee recommends a comprehensive, integrated health care system for all astronauts, active (including those in training to be astronauts) and nonactive (including those who are retired), and their immediate families. A model similar to that of the military, but with modifications, is outlined below. The rationale for such a comprehensive system, in the committee's view, is based on the degree of risk of long-duration missions to both current and future astronauts and on the practical difficulty of obtaining sufficient clinical data to ensure a reasonable degree of safety for such missions. Comprehensiveness maximizes data collection and tends to diminish risk by promoting solutions. Care for the family is justified both for the reasons defined by the military—for example, as a benefit that augments low pay and that is a necessity in areas where health care may otherwise be unavailable and in a profession where personnel are often subject to geographical moves—and because of the interrelationship between family health (both physical and psychosocial) and astronaut health during the stressful separation while the astronaut is on long-duration space missions.

Comprehensiveness means that all health care for the astronaut is, at a minimum, coordinated through the astronaut health care program and covers all periods while the astronaut is active, including the selection, premission, intramission, postmission, and intermission phases. For the retired astronaut, health care would continue to be available through the astronaut health care program as long as the astronaut chooses. However, participation in NASA's Longitudinal Study of Astronaut Health should be strongly encouraged by providing explanations about the long-term advantage to the individual astronaut and to those who will follow. Comprehensiveness also means that clinical data will be retrospectively as well as prospectively collected and analyzed. The committee believes that much of relevance can be learned from retrospective review of medical records, mission debriefings, interviews, and stored clinical samples.

A health care system for astronauts should also be integrated. That is, it should include, in addition to a care component, a research and a training component. What is learned from basic and clinical research should be integrated into care plans, and what is learned in the process of care, if relevant, should stimulate research. In addition, since most forms of health promotion and disease and injury prevention depend on the individual, the astronaut should be trained not only in preventive measures but also in routine medical care. Currently, NASA coordinates these functions in its Clinical Care Capability Development Project (NASA, 1998b). This would appear to be an appropriate initiative, but more may be needed, as described below.

Standard of Care

The committee believes that the standard of clinical care for a health care system for astronauts should be equivalent to that of the best terrestrial health care system for those problems that occur pre- and postmission and in the intermission phase. The committee is using the best possible medical practice as the definition of standard of care rather than using "standard of care" in the legal term-of-art sense. Standard of care in most states is defined by common law and case law, not by statute. For care during the mission, the standard of care must, of necessity, evolve. For example, should a long-duration mission beyond Earth orbit occur in the near future, the standard would likely be less ideal than it would be if the mission occurred farther in the future, when new knowledge and the further development of prevention and treatment measures, care protocols, and skills could be incorporated. The goals of the care system should be to maximize the individual astronaut's ability to function as a productive member of the crew during the mission and to maintain or to restore normal function during and upon completion of the mission.

Medical Care

NASA (NASA, 1998a) has outlined a strategy for health care on long-duration missions with which the committee's principles are consonant. The components of that strategy are listed below:

Preflight

- Determine the expected frequencies of health conditions that will require medical intervention.
- Institute preventive measures.
- Match the severity of the health condition with the treatment necessary to cure or control the condition.

In Flight

- Determine the level of skill necessary on board to diagnose and treat the expected health conditions.
- Determine the necessary diagnostic technologies (kits, informatics).
- Determine the necessary treatment technologies (drugs, biologics, supplies).

- Develop measures to stabilize a crewmember if cure or control is not possible.
- Develop a plan for disposition of the crewmember in case of death.

Postflight

- Plan for long-term care of chronic health problems.
- Periodically review the strategy.

The committee accepts the components listed above with the following caveats. NASA has implemented the Critical Path Roadmap project to determine the health conditions requiring medical intervention. The project initially identified 11 "discipline risk areas": advanced life support; bone loss; cardiovascular alterations; environmental health; food and nutrition; human performance; immunology, infection, and hematology; muscle alterations and atrophy; neurovestibular adaptation; radiation effects; and space medicine. Chapters 2 and 3 of this report address what is known in these fields and how issues in these fields may be approached. Once identified, NASA's approach to a problem is to develop countermeasures. The committee has concern that use of the concept of countermeasures leads to technological solutions to biological problems. Instead, the committee urges that for a biological problem consideration be given to understanding the fundamental processes leading to the biological problem. Causation can then be ascertained from this understanding, and only then can a solution be defined.

The committee wishes to emphasize an additional aspect of the preflight phase, that of prevention. Primary prevention begins with the proper selection and training of astronauts and astronaut teams for missions (a subject dealt with in detail in Chapter 5 of this report) and the use of engineering design. NASA should continue its validation of these processes to maintain the confidence of the astronaut corps, external bodies, and the public. Secondary prevention is dependent on proper safety standards and good strategies protective against occupational hazards. Both are, in turn, dependent on knowledge of risk. Tertiary prevention is dependent on having well-trained medical personnel and databases on board the spacecraft and access to qualified personnel and other resources on the ground.

An additional point of importance is that of triage, which, to some degree, will govern care during long-duration missions beyond Earth orbit. Historically, triage involves making decisions to maximize the benefits of care to the group, given the resources available to the group. The committee anticipates that care will be initiated to the maximum available, assuming that planning for the mission has included a level of appropriate diagnostic and therapeutic technologies and materials. (Evidence from the Antarctic experience indicates that the ingenuity of humans enables heroic treatments even beyond that considered possible with the available medical materials.) Initially, the triage principle will come into play in the preflight phase, when decisions are made about the intramission resources that will be available. Other instances in which triage may be invoked are that of a catastrophe involving crewmembers and that involving a decision to withdraw care. Both will require preparation preflight, knowing with certainty that not every possible scenario can be anticipated. Although it is theoretically possible that a planned long-duration mission beyond Earth orbit could be aborted because of the medical condition of a crewmember, the committee considers this unlikely because of the practical difficulty of doing so. Nevertheless, it is valuable to develop guidelines prospectively to address this issue. Medical condition data from the nuclear submarine service could provide the conceptual framework for such guidelines.

Because of the relative paucity of data on the long-term effects on health of long-duration spaceflight, NASA has been unable to predict which, if any, long-term chronic effects will need to be addressed. Use of the ISS as a clinical research platform should help with this endeavor. Initial evidence suggests several areas, some with short-term effects and several with longer-term effects, that may require attention. These include dehydration, neurosensory motor control adjustment, decreased muscle strength and endurance, diminished cardiac reserve, decreased immune response, regional osteopenia, increased risk of renal calculi, and increased risk of radiation-induced cancer. To address these issues, long-term data collection and analysis through strict application of the Longitudinal Study of Astronaut Health are critical. As more is learned from the experiences of astronauts aboard the ISS, standards can be developed on what is normal or abnormal during a space mission as well as what is an acceptable postmission status. Such standards can direct treatment and rehabilitation efforts critical to longer missions.

Medical Informatics

In a way similar to that in which new techniques and technologies such as laparoscopy, microtechniques, and videoscopy have transformed the skills required for surgery, medical informatics will transform the medical skills required for diagnosis and treatment, perhaps to an even greater extent. Medical informatics refers to information systems that support prevention, monitoring, diagnosis, decision support, intervention, and treatment. The key elements of such a system are user interfaces and displays, artificial intelligence, automation, smart systems, sensors, data acquisition systems including handheld computer devices, computer-based training, and simulation and a communication infrastructure (NASA, 1999).

NASA is to be commended for having gone beyond simple communication to telemedicine (or medicine at a distance) to exploring the larger role of medical informatics in the space program. Although this conceptual leap and its practical implications will be helpful on the ISS, medical informatics will be a necessity on any long-duration mission beyond Earth orbit. NASA has reached out to the best minds and most experienced people in the field (NASA, 1999) and continues to solicit their help. NASA is working with academia and industry in the form of Meditac, a commercial space center, to carry out applied research and to develop products, initially in telemedicine and, increasingly, in informatics. An example of the output of this collaboration was the development and testing in an analog environment (a 1996 Mt. Everest expedition) of lightweight core body temperature sensors. NASA has developed and tested numerous telemedicine technologies in low Earth orbit. These include the provision of an advanced cardiac life support system on Neurolab and STS-90 and the Telemedicine Instrumentation Pack (in conjunction with Meditac), which is a computer-based system that can videoimage the eye, ear, nose, throat, and skin and that may be used to monitor the electrocardiogram, systemic oxygen pressure, blood pressure, and heart rate.

Medical informatics can provide on board the spacecraft the information base that will underlie medical decision making during long-duration space missions. Telemedicine will provide the consultative link to Earth-bound systems of ground crews, flight surgeons, and consultants, albeit with inherent time delays. In a broader sense, medical informatics is a conceptual shift that views the physician not as an experience-based practitioner but as an information manager willing to use guidelines to provide care. Additionally, medical informatics will provide sophisticated techniques for the development of data-based guidelines for care and technologies for the storage, linkage, and rapid retrieval of information including that related to the new and rapidly advancing discipline of medical genomics.

Important for the entire crew of a long-duration space mission, but specifically for those responsible for medical care on board the spacecraft, is the capability for continuing education and updating of skills. For the medical professional on a long-duration mission, the degradations in unused knowledge and technical skills can be substantial. Medical informatics is thus necessary not only for knowledge and skills to be applied at a particular moment but also for continuing competence. Finally, since information will continue to develop while a long-duration mission is ongoing, there must be a mechanism for continued updating of onboard software and databases via linkage to the ground.

Personnel and Training

As stated earlier, the salient feature of a health care system for astronauts is comprehensiveness. In this context, comprehensiveness does not mean that every health care service is supplied at every site only by an individual employed full time within the system. Instead, it means that necessary and appropriate services be available. Because of the relatively small size of the astronaut corps, as well as the need for some degree of individual choice of practitioner by astronauts, this requires an extensive consultant and specialty capacity external to the primary system, as well as a means to attract, maintain, and promote clinical capability within the primary system.

There is, however, an additional consideration for space medicine consultation. That is the likelihood that the need for consultation by those on board the spacecraft will likely be sudden and urgent. NASA has traditionally relied on the telephone and electronic mail to contact its outside medical consultants; it recognizes that more real-time technologies must be used. Given the separation in time and space of the astronauts on a long-duration mission from their Earth-bound colleagues, a delay in urgent consultation would be detrimental. It will be important to incorporate continuing and future developments in medical informatics at all levels to ensure future success during travel beyond Earth orbit.

Issues of Capability

The committee's perspective causes it to have concern about three issues related to the clinical capability within the primary health care system: the small *n* problem, the flight surgeon, and continuing competence.

Small *n* Problem The committee's first concern is the small *n* problem (referring in this case to the relatively small numbers of individuals in the astronaut corps) and its effect on attracting good practitioners and maintaining their skills. NASA staff are well aware of this problem. Physically connecting the employee and the occupational health clinic at JSC with the astronaut clinic represents an attempt to address this problem.

Flight Surgeon The committee heard from a number of flight surgeons, all of who impressed the committee with their dedication to the space program. At the same time, their status within the program seems less than it should or could be. The committee believes, because of the evidence that it heard and reviewed, that the role of the flight surgeon as counselor and primary care physician is a key one. One of the most important reasons for this is the isolation experienced by the astronaut during space travel and the lack of privacy that demands the availability of a confidential channel of information. When the information is both sensitive for the individual and potentially critical for crew performance, a relationship of mutual trust and respect is necessary. The committee believes that it is important that NASA continue to examine ways to evaluate and upgrade the status and role of the flight surgeon, including ways to make the practice environment attractive.

Continuing Competence Medical professionals, especially physicians and nurses, put great store in continuing education to upgrade their knowledge and skills. In settings such as private practice and academia, there are multiple formal and informal structures and opportunities to interact with colleagues and to study. In the isolated setting in which most clinicians at NASA find themselves, few of these opportunities exist. Again, NASA staff recognize the problem. The committee encourages NASA space medical staff, in concert with those at the highest levels at NASA, to continue to seek approaches to provide ways for clinical staff, flight surgeons, and physician-astronauts to maintain and upgrade their knowledge and skills.

Approaching the Issues of Capability

From the committee's perspective, one way of approaching the issues of capability is this: medical professionals, especially physicians, enjoy diversity. The role of the flight surgeon is somewhat diverse, in that the individual serves both as the primary contact for the astronaut while the astronaut is on a mission and as a clinic physician. However, to enhance the role of the flight surgeon, to upgrade competences, and to maintain interest, all of which will increase the ability of NASA to attract good people, more needs to be done. The flight surgeon could be increasingly involved in the preflight training of the crew, not only in relation to in-flight issues but also in relation to health promotion throughout the continuum of the premission, intramission, and postmission phases. Furthermore, since the health care system for astronauts must evolve over the next several decades, clinicians with management expertise may be attracted to the task of building such a system. Lastly, since the committee strongly recommends a clinical research strategy as the basis for ensuring the health and safety of the astronaut corps, flight surgeons may have a role in designing the research and its application. The model that the committee suggests is that of the academic physician, who is active in multiple components of health care, not only in the acute-care phase. By reconceptualizing the role of the flight surgeon to that of a substantial component of the space program team, the committee believes that the health of the astronauts will be enhanced.

Finally, current standards of practice and expectations by the medical community and the general public indicate that a physician should be part of the crew. Although in the analog environment of long-duration submarine missions no physician is generally included as part of the crew, the committee believes there are two distinctions that argue, at least initially, for a physician to be on board the spacecraft during a long-duration mission beyond Earth orbit. Unlike NASA, the U.S. Navy has had substantial experience with having a highly trained medical corpsman as the principal clinician on board a submarine. Furthermore, the time frame is distinctly different for travel beyond Earth orbit: telecommunication can be virtually instantaneous with a submarine, and when necessary, evacuation can and has taken place; that will not be the case in deep space.

Evaluation and Performance Improvement

The committee recommends that NASA establish a comprehensive health care system for astronauts that is integrated with other components of the space program and informed by the best knowledge base and information system available. Strong and sophisticated quality measurement and performance improvement methodologies are essential components of such a system. The science of quality measurement and performance improvement has improved greatly over the past decade as quality measurement and performance improvement methodologies have developed, as the public has grown to expect more from its health care system, and as medical professionals have grown to understand the value in objectively evaluating the outcomes of care. A report of the President's Advisory Commission on Consumer Protection and Quality in the Health Care Industry (1998), the Institute of Medicine publication *To Err is Human: Building a Safer Health System* (IOM, 2000), and related reports (IOM 2001a,b), as well as the public's response to them, indicate that quality improvement in health care is imperative. The Institute of Medicine has been and continues to be at the forefront of public and professional activity in the field, for example, as sponsor of the National Roundtable on Health Care Quality (1996-1998) and the Quality of Health Care in America project (1998 and ongoing). NASA's involvement in the latter and the National Quality Forum would enable NASA to obtain the best national and international advice from the field.

Advances in quality measurement include the following (IOM, 1999b):

- automated ways of reminding practitioners about the appropriate use of medications and the creation of a database about diseases and their treatments;
- the measurement of risk-adjusted mortality and investigation of the science and art of adjusting the measured outcomes of care, taking into account the severity of a patient's illness and other risk factors such as the presence of other health conditions;
- measuring errors that occur in organizations so that organizations can pinpoint how such errors occur and how they can be prevented;
- the development of patient-reported measures of quality that allow organizations to compare a patient's experience with the patient's expectations;
- quality measurement in integrated delivery systems that include multiple settings of care; and
- the translation of well-developed clinical practice guidelines into performance measures.

These examples represent ways in which NASA might go about evaluating the results of a health care system for astronauts. Moreover, the use of state-of-the-art methodologies of quality measurement and performance improvement would make a significant statement that, when NASA is in fact ready to send humans deep into space, astronaut health and safety would be its top priority.

CONCLUSION AND RECOMMENDATION

Conclusion

Crew health has not received the attention that it must receive to ensure the safety of astronauts on long-duration missions beyond Earth orbit, nor has NASA sufficiently integrated astronaut health care into mission operations.

- *Currently, there is no current comprehensive and inclusive strategy to provide optimum health care for astronauts on long-duration missions beyond Earth orbit, nor is there sufficient coordination of health care needs with the engineering aspects of such missions.*
- *An effective health care system is founded on data that are accumulated, analyzed, and used to continuously improve health care for astronauts on future space missions. Inherent in an appropriate health care system is a mechanism that can be used to gather and analyze data relevant to key variables. NASA could have collected and analyzed many more medical data had a comprehensive health care system focused on astronauts been in place and been given the priority and resources that it needed.*
- *Although the equipment and expertise that will be needed to provide health care during future long-duration missions beyond Earth orbit cannot be reliably predicted, a health care system that is data driven and linked to a research strategy will position NASA to better monitor pertinent developments and meet future challenges.*

Recommendation

NASA should develop a comprehensive health care system for astronauts for the purpose of collecting and analyzing data while providing the full continuum of health care to ensure astronaut health. A NASA-sponsored health care system for astronauts should

- **care for current astronauts, astronauts who are in training, and former astronauts, as well as, where appropriate, their families;**
- **cover all premission, intramission, and postmission aspects of space travel;**
- **incorporate innovative technologies and practices—including clinical practice guidelines—into prevention, diagnosis, treatment, and rehabilitation, including provision for medical care during catastrophic events and their sequelae;**

- **be uniform across the international space community and cooperatively developed with the international space community; and**
- **receive external oversight and guidance from prominent experts in clinical medicine.**

References

AAMC (Association of American Medical Colleges). 1999. *Breaking the Scientific Bottleneck. Clinical Research: A National Call to Action.* Report of the Clinical Research Summit. Washington, D.C.: Association of American Medical Colleges.

AAOS (American Academy of Orthopaedic Surgeons). 1996. *New Bone Paste Heals Fractures Faster Than Traditional Methods.* AAOS Academy News. Online. Available at www.aaos.org/wordhtml/aaosnews/paste.htm. Accessed February 2001.

ACS (American College of Surgeons). 1999. *Resources for the Optimal Care of the Injured Patient.* Chicago: American College of Surgeons.

Agathon, M. 1998. Part of education in behaviour. Therapy approaches of anxiety disorders. *Annales Medico-Psychologiques* 156(10):697–699.

AHA (American Heart Association). 2000. ECC Guidelines, Part 2. Ethical Aspects of CPR and ECC. *Circulation* 102:1–12.

Alfrey, C. P., M. M. Udden, C. S. Leach-Huntoon, T. Driscoll, and M. H. Pickett. 1996. Control of red blood cell mass in space flight. *Journal of Applied Physiology* 81(1):98–104.

Alkire, M. T., R. J. Haier, J. J. Fallon, and L. Cahill. 1998. Hippocampal, but not amygdala, activity at encoding correlates with long-term, free recall of nonemotional information. *Proceedings of the National Academy of Sciences* 95:14506–14510.

American Cancer Society. 2000. *Cancer Facts and Figures.* Atlanta: American Cancer Society.

Arbeille, P., G. Gauquelin, J. M. Pottier, L. Pourcelot, A. Guell, and C. Gharib. 1992. Results of a 4-week head-down spaceflight. *Aviation, Space and Environmental Medicine* 63(1):9–13.

Arbeille, P., F. Achaibou, G. Fomina, J. M. Pottier, and M. Porcher. 1996. Regional blood flow in microgravity: adaptation and deconditioning. *Medicine and Science in Sports and Exercise* 28(10 Suppl.):S70–S79.

Athma, P., R. Rappaport, and M. Swift. 1996. Molecular genotyping shows that ataxia-telangiectasia heterozygotes are predisposed to breast cancer. *Cancer Genetics and Cytogenetics* 92(2):130–134.

ATSDR (Agency for Toxic Substances and Disease Registry). 1991. *Medical Guidelines for Acute Chemical Exposures, Managing Hazardous Materials Incidents.* Atlanta: ATSDR.

ATSDR. 1997. *Hazardous Substances Emergency Events Surveillance Annual Report.* Atlanta: ASTDR.

Bachman, W. 1988. Nice guys finish first: a SYMLOG analysis of U.S. Naval Commands. *In: The SYMLOG Practitioner: Applications of Small Group Research.* R. B. Polley, A. P. Hare, and J. P. Stone, eds. New York: Praeger.

Bagian, J. P., and D. F. Ward. 1994. A retrospective study of promethazine and its failure to produce the expected incidence of sedation during space flight. *Journal of Clinical Pharmacology* 34:649–651.

Baibekov, I. M., R. S. Mavlyan-Khodzhaev, V. P. Tumanov, and K. K. Usmanov. 1995. Effects of low-intensity infrared laser radiation on the healing of dermatomal wounds. *Byulleten' Eksperimental'Noi Biologii I Meditsiny* 119(2):218–224.

Baisch, F. J. 1993. Body fluid distribution in man in space and effect of lower body negative pressure treatment. *Clinical Investigation* 71(9):690–699.

Baker, E. L. 1988. Surveillance of occupational disease: strategies for improving physician recognition and reporting, pp. 79–80. *In: Role of the Primary Care Physician in Occupational and Environmental Medicine.* Washington, D.C.: National Academy Press.

Bales, R. F. 1999. *Social Interaction Systems: Theory and Measurement.* New Brunswick, N.J.: Transaction Publishers.

Barger, L. K., and C. A. Czeisler. 2000. Research letter submitted to the Institute of Medicine Committee on Creating a Vision for Space Medicine During Travel Beyond Earth Orbit, July 7.

Barlow, D. 1996. Health care policy, psychotherapy research and the future of psychotherapy. *American Psychologist* 51:1050–1058.

Beck, A. T. 1993. Cognitive therapy: past, present and future. *Journal of Consulting and Clinical Psychology* 61:194–198.

Beck, A. T., and G. Emery. 1985. *Anxiety Disorders and Phobias: A Cognitive Perspective.* New York: Basic Books.

Benichou, J., C. Byrne, and M. Gail. 1997. An approach to estimating exposure-specific rates of breast cancer from a two-stage case control study within a cohort. *Statistics in Medicine* 16(1–3):133–151.

Berry, C. A. 1969. Preliminary clinical report of the medical aspects of Apollos VII and VIII. *Aerospace Medicine* 40:245–254.

Biglan, A. 1995. *Changing Cultural Practices: A Contextualist Framework for Intervention Research.* Reno, NV: Context Press.

Billica, R. 2000. In-flight Medical Events for US Astronauts During Space Shuttle Program STS-1 through STS-89, April 1981–January 1998. Presentation to the

Institute of Medicine Committee on Creating a Vision for Space Medicine During Travel Beyond Earth Orbit, February 22, Johnson Space Center, Houston.

Billica, R. D., S. C. Simmons, K. L. Mathes, B. A. McKinley, C. C. Chuang, M. L. Wear, and P. B. Hamm. 1996. Perception of the medical risk of spaceflight. *Aviation, Space and Environmental Medicine* 67(5):467–473.

Birketvedt, G. S., J. Florholmen, J. Sundsfjord, B. Osterud, D. Dinges, W. Bilker, and A. Stunkard. 1999. Behavioral and neuroendocrine characteristics of the night-eating syndrome. *Journal of the American Medical Association* 282(7):657–663.

Biros, M. H., R. J. Lewis, C. M. Olson, J. W. Runge, R. O. Cummins, and N. Fost. 1995. Informed consent in emergency research: consensus statement from the Coalition Conference of Acute Resuscitation and Critical Care Researchers. *Journal of the American Medical Association* 273(16):1283–1287.

Biros, M. H., J. W. Runge, R. J. Lewis, and C. Doherty. 1998. Emergency medicine and the development of the Food and Drug Administration's final rule on informed consent and waiver of informed consent in emergency research circumstances. *Academic Emergency Medicine* 5(4):359–368.

Black, F. O., S. W. Wade, and A. Arshi. 1999. Roll-tilt perception using a somatosensory bar task. *In: Proceedings of the First Biennial Space Biomedical Investigators' Workshop*, January 11–13. Houston: National Aeronautics and Space Administration.

Boal, K. B., and R. Hooijberg. 2000. Strategic leadership research: moving on. *Leadership Quarterly* 11(4):515–549.

Bode, P. J., M. J. Edwards, M. C. Kruit, and A. B. Van Vugt. 1999. Sonography in a clinical algorithm for early evaluation of 1671 patients with blunt abdominal trauma. *American Journal of Roentgenology* 172:905–911.

Bold, R. J., P. F. Mansfield, D. H. Berger, R. E. Pollock, S. E. Singletary, F. C. Ames, C. M. Balch, D. C. Hohn, and M. I. Ross. 1998. Prospective, randomized, double-blind study of prophylactic antibiotics in axillary lymph node dissection. *American Journal of Surgery* 176(3):239–243.

Borman, W. C., M. A. Hanson, and J. W. Hedge. 1997. Personnel selection. *Annual Review of Psychology* 48:299–337.

Borst, C. 2000. Operating on a beating heart. *Scientific American* 283(4):58–63.

Brady, J. 1990. Toward applied behavior analysis of life aloft. *Behavioral Science* 35:11–23.

Brady, J., and M. Anderson. 1991. Small groups in confined microsocieties. *In: From Antarctica to Outer Space: Life in Isolation and Confinement.* A. Harrison, Y. Clearwater, and C. McKay, eds. New York: Springer-Verlag.

Brady, J., T. Kelly, and R. Hienz. 1999. Stability and precision of performance during space flight. *In: Proceedings of the First Biennial Space Biomedical Investigators' Workshop*, January 11–13. Houston: National Aeronautics and Space Administration.

Braunwald, E., ed. 1999. *Heart Disease.* New York: Harcourt Brace.

Buckey, J. C., Jr., F. A. Gaffney, L. D. Lane, B. D. Levine, D. E. Watenpaugh, S. J. Wright, C. W. Yancy, Jr., D. M. Meyer, and C. G. Blomqvist. 1996a. Central venous pressure in space. *Journal of Applied Physiology* 81(1):19–25.

Buckey, J. C., Jr., L. D. Lane, B. D. Levine, D. E. Watenpaugh, S. J. Wright, W. E. Moore,

F. A. Gaffney, and C. G. Blomqvist. 1996b. Orthostatic intolerance after spaceflight. *Journal of Applied Physiology* 81(1):7–18.

Bungo, M. W., J. B. Charles, and P. C. Johnson, Jr. 1985. Cardiovascular deconditioning during spaceflight and the use of saline as a countermeasure to orthostatic intolerance. *Aviation, Space and Environmental Medicine* 56(10):985–990.

Burrough, B. 1998. *Dragonfly: NASA and the Crisis Aboard Mir*. New York: HarperCollins Publishers.

Burrows, W. E. 1998. *This New Ocean*. New York: Random House.

Cacioppo, J. T., and L. G. Tassinary. 1990. *Principles of Psychophysiology: Physical, Social, and Inferential Elements*. New York: Cambridge University Press.

Cahill, L., R. J. Haier, J. Fallon, M. T. Alkire, C. Tang, D. Keator, J. Wu, and J. L. McGaugh. 1996. Amygdala activity at encoding correlated with long-term, free recall of emotional information. *Proceedings of the National Academy of Sciences* 93:8016–8021.

Caillot-Augusseau, A., M. H. Lafage-Proust, C. Soler, J. Pernod, F. Dubois, and C. Alexandre. 1998. Bone formation and resorption biological markers in cosmonauts during and after a 180-day spaceflight (Euromir 95). *Clinical Chemistry* 44(3):578–585.

Campbell, M. R. 1999. Surgical care in space. *Aviation, Space and Environmental Medicine* 70(2):181–184.

Campbell, M. R., and R. D. Billica. 1992. A review of microgravity surgical investigations. *Aviation, Space and Environmental Medicine* 63(6):524–528.

Campbell, M. R., R. D. Billica, and S. L. Johnston III. 1993. Surgical bleeding in microgravity. *Surgery, Gynecology and Obstetrics* 177(2):121–125.

Campbell, M. R., R. D. Billica, R. Jennings, and S. L. Johnston III. 1996. Laparoscopic surgery in weightlessness. *Surgical Endoscopy* 10(2):111–117.

Caplan, R. A., J. L. Benumof, F. A. Berry, C. D. Blitt, R. H. Bode, F, W. Cheney, R. T. Connis, O. F. Guidry, and A. Ovassapian. 1993. Practice guidelines for management of the difficult airway. A report by the American Society of Anesthesiologists Task Force on Management of the Difficult Airway. *Anesthesiology* 78(3):597–602.

Cautela, J. R., and M. Ishaq. 1996. *Contemporary Issues in Behavior Therapy: Improving the Human Condition*. New York: Plenum.

Cavanagh, P. R., B. L. Davis, and T. A. Miller. 1992. A biomechanical perspective on exercise countermeasures for long-term spaceflight. *Aviation, Space and Environmental Medicine* 63(6):482–485.

Charles, J. 2000. Critical Path for Countermeasure Development. Presentation to the Institute of Medicine Committee on Creating a Vision for Space Medicine During Travel Beyond Earth Orbit, February 23, Johnson Space Center, Houston.

Charles, J. B., and C. M. Lathers. 1994. Summary of lower body negative pressure experiments during spaceflight. *Journal of Clinical Pharmacology* 34(16):571–583.

Charles, J. B., J. M. Fritsch-Yelle, P. A. Whitson, M. L. Wood, T. E. Brown, and G. W. Fortner. 1999. Cardiovascular deconditioning. *In*: *Extended Duration Orbiter Medical Project*. Final Report 1989–1995, NASA Special Publication SP-1999-534. C. F. Sawin, G. R. Taylor, and W. L. Smith, eds. Houston: National Aeronautics and Space Administration.

Cintron, N. M., L. Putcha, Y. M. Chen, and J. M. Vanderploeg. 1987. Inflight salivary pharmacokinetics of scopolamine and dextroamphetamine. *In: Results of the Life*

Sciences DSOs Conducted Aboard the Space Shuttle 1981–1986. M. W. Bongo, T. M. Bagian, M. A. Bowman, and B. M. Levitan, eds. Houston: Johnson Space Center, National Aeronautics and Space Administration.

CIOMS (Council for International Organizations of Medical Sciences). 1993. Ethics and Research on Human Subjects: International Guidelines. *In: Proceedings of the XXVI CIOMS Conference*, Geneva, Switzerland, February 5–7, 1992. Z. Bankowski and R. J. Levine, eds. Geneva: Council for International Organizations of Medical Sciences.

Claus, A. B., J. M. Schildkraut, W. D. Thompson, and N. J. Risch. 1996. The genetic attributable risk of breast and ovarian cancer. *Cancer* 77(11):2318–2324.

Cohen, M. M., S. M. Ebenholtz, and B. J. Linder. 1995. Effects of optical pitch on oculomotor control and the perception of target elevation. *Perception and Psychophysics* 57(4):433–440.

Collins, M. 1974. *Carrying the Fire: An Astronaut's Journeys*. New York: Farrar, Straus and Giroux.

Collins, M. 1990. *Mission to Mars*. New York: Grove Weidenfeld.

Colvard, M., P. Kuo, R. Caleel, J. Labo, and R. Self. 1992. Laser surgery procedures in the operational KC-135E aviation environment. *Aviation, Space and Environmental Medicine* 63(7):619–623.

Convertino, V. A. 1996a. Clinical aspects of the control of plasma volume at microgravity and during return to one gravity. *Medicine and Science in Sports and Exercise* 28(10 Suppl.):S45–S52.

Convertino, V. A. 1996b. Exercise as a countermeasure for a physiological adaptation to prolonged spaceflight. *Medicine and Science in Sports and Exercise* 28(8):999–1014.

Convertino, V. A., J. L. Polet, K. A. Engelke, G. W. Hoffler, L. D. Lane, and C. G. Blomqvist. 1997. Evidence for increased β-adrenoreceptor responsiveness induced by 14 days of simulated microgravity in humans. *American Journal of Physiology* 273(1 Part 2):R93–R99.

Cullen, M. R., and L. Rosenstock. 1994. Principles and practice of occupational medicine. *In: Textbook of Clinical Occupational and Environmental Medicine*. L. Rosenstock and M. R. Cullen, eds. Philadelphia: W. B. Saunders.

Czeisler, C. A., and K. P. Wright, Jr. 1999. Influence of light on circadian rhythmicity in humans. *In: Regulation of Sleep and Circadian Rhythms*. F. W. Turek and P. C. Zee, eds. New York: Marcel Dekker.

Czeisler, C. A., M. P. Johnson, J. F. Duffy, E. N. Brown, J. M. Ronda, and R. E. Kronauer. 1990. Exposure to bright light and darkness to treat physiologic maladaptation to night work. *New England Journal of Medicine* 322:1253–1259.

D'Agostino, H. B. 1999. Transcatheter fluid drainage. *In: Vascular and Interventional Radiology*. K. Valji, ed. Philadelphia: W. B. Saunders.

Damasio, A. B. 1994. *Descartes' Error*. New York: Avon.

Deaver, D. R., R. P. Amann, R. H. Hammerstedt, R. Ball, D. N. Veeamachaneni, and X. J. Musacchia. 1992. Effects of caudal elevation on testicular function in rats. Separation of effects on spermatogenesis and steroidogenesis. *Journal of Andrology* 13(3):224–231.

Derendorf, H. 1994. Pharmacokinetic/pharmacodynamic consequences of space flight. *Journal of Clinical Pharmacology* 34:684–691.

DHHS (U.S. Department of Health and Human Services). 1992. *Model Trauma Care System Plan*. Washington, D.C.: U.S. Department of Health and Human Services.

DiGioia, A. M., B. D. Colgan, and N. Koerbel. 1998. Computer-aided surgery. *In: Cybersurgery*. R. M. Satava, ed. New York: Wiley-Liss.

Dijk, D. J., J. F. Duffy, and C. A. Czeisler. 1992. Circadian and sleep/wake dependent aspects of subjective alertness and cognitive performance. *Journal of Sleep Resolution* 1(2):112–117.

Dinges, D., and H. Van Dongen. 1999. Countermeasures to neurobehavioral deficits from cumulative partial sleep deprivation during spaceflight. *In: Proceedings of the First Biennial Space Biomedical Investigators' Workshop*, January 11–13. Houston: National Aeronautics and Space Administration.

Dinges, D. F., F. Pack, K. Williams, K. A. Gillen, J. W. Powell, G. E. Ott, C. Aptowicz, and A. I. Pack. 1997. Cumulative sleepiness, mood disturbance, and psychomotor vigilance performance decrements during a week of sleep restricted to 4–5 hours per night. *Sleep* 20(4):267–277.

Dlugos, D. J., P. L. Perrotta, and W. G. Horn. 1995. Effects of the submarine environment on renal-stone risk factors and vitamin D metabolism. *Undersea and Hyperbaric Medicine* 22(2):145–152.

Doll, R. E., and E. K. E. Gunderson, 1970. The relative importance of selected behavioral characteristics of group members in an extreme environment. *Journal of Psychology* 75:23-237.

Duffy, L. 1993. Team decision-making biases: an information-processing perspective. *In: Decision Making in Action: Models and Methods*. G. Llein, J. Orasanu, R. Calderwood, and C. Zsambok, eds. Norwood, N.J.: Ablex Publishing Corporation.

Easton, D. F., D. Ford, and T. Bishop. 1995. Breast Cancer Linkage Consortium: breast and ovarian cancer incidence in BRCA 1-mutation carriers. *American Journal of Human Genetics* 56(1):265–271.

Eddy, D., S. Schiflett, R. Schlegel, and R. Shehab. 1999. Cognitive performance in seven shuttle astronauts. *In: Proceedings of the First Biennial Space Biomedical Investigators' Workshop*, January 11–13. Houston: National Aeronautics and Space Administration.

Edgerton, V. R., M. Y. Zhou, Y. Ohira, H. Klitgaard, B. Jiang, G. Bell, B. Harris, B. Saltin, P. D. Golnick, R. R. Roy, M. K. Day, and M. Greenisen. 1995. Human fiber size and enzymatic properties after 5 and 11 days of spaceflight. *Journal of Applied Physiology* 78(5):1733–1739.

Emurian, H., J. Brady, J. Meyerhoff, and E. Mougey. 1981. Behavioral and biological interactions with confined microsocieties in a programmed environment. *In: Space Manufacturing*. New York: American Institute of Aeronautics and Astronautics.

Estupiñán-Day, S. 2000. Atraumatic Restorative Treatment (ART): A Potentially Useful Approach to Managing Dental Lesions in Space. Presentation to the Institute of Medicine Committee on Creating a Vision for Space Medicine During Travel Beyond Earth Orbit, April 17, National Academy of Sciences, Washington, D.C.

Fehr, T. G. 1996. Relevant therapeutic effects of transcendental meditation. *Psychotherapie, Psychosomatik, Medizinische Psychologie* 46(5):178–188.

Feiveson, A. H. 2000. Quantitative Assessment of Countermeasure Efficacy for Long-Term Space Missions. Presentation to the Institute of Medicine Committee on Strategies for Small Number Participant Clinical Research Trials, September 28, National Academy of Sciences, Washington, D.C.

Ferguson, J. K. 1977. Flight crew health stabilization program. *In: The Apollo-Soyuz Test*

Project Medical Report. A. E. Nicogossian, ed. NASA Special Publication 411. Washington, D.C.: National Aeronautics and Space Administration.

Ferguson, J. K., G. W. McCollum, and B. L. Portnoy. 1977. Analysis of the Skylab flight crew health stabilization program. *In: Biomedical Results from Skylab.* R. S. Johnston and L. F. Dietlein, eds. NASA Special Publication 377. Washington, D.C.: National Aeronautics and Space Administration.

Fisher, C. 2000. Clinical Care Capability Development Program. Presentation to the Institute of Medicine Committee on Creating a Vision for Space Medicine During Travel Beyond Earth Orbit, February 23, Johnson Space Center, Houston.

Fisher, R., E. Kopelman, and A. K. Schneider. 1994. *Beyond Machiavelli: Tools for Coping with Conflict.* Cambridge, Mass.: Harvard University Press.

Fletcher, L. D. 1983. Dental observations at Australian Antarctic Stations. *Australian Dental Journal* 28(5):281–285.

Flynn, C., and A. Holland. 2000. Behavioral Health and Performance. Presentation to the Institute of Medicine Committee on Creating a Vision for Space Medicine During Travel Beyond Earth Orbit, February 22, Johnson Space Center, Houston.

Ford, D., D. F. Easton, M. Stratton, S. Narod, D. Goldgar, P. Devilee, D. T. Bishop, B. Weber, G. Lenoir, J. Chang-Claude, H. Sobol, M. D. Teare, J. Struewing, A. Arason, S. Schernek, J. Peto, T. R. Rebbeck, P. Tonin, S. Neuhausen, R. Barkardottir, J. Eyfjord, H. Lynch, B. A. Ponder, S. A. Gayther, and M. Zelada-Hedman. 1998. Genetic heterogeneity and penetrance analysis of the BRCA 1 and BRCA 2 genes in breast cancer families. *American Journal of Human Genetics* 62(3):676–689.

Fortney, S. M. 1991. Development of lower body negative pressure as a countermeasure for orthostatic intolerance. *Journal of Clinical Pharmacology* 31(10):888–892.

Frazer, T. 1968. Leisure and recreation in long duration space missions. *Human Factors* 10:483–488.

Frazier, S. H. 1968. Comprehensive management of psychiatric emergencies. *Psychosomatics* 9(1):7–11.

Fritsch-Yelle, J. M., J. B. Charles, M. M. Jones, L. A. Beightol, and D. L. Eckberg. 1994. Spaceflight alters autonomic regulation of arterial pressure in humans. *Journal of Applied Physiology* 77(4):1776–1783.

Fritsch-Yelle, J. M., P. A. Whitson, R. L. Bondar, and T. E. Brown. 1996. Subnormal norepinephrine release relates to presyncope in astronauts after spaceflight. *Journal of Applied Physiology* 81(5):2134–2141.

Gail, M. H., L. A. Brinton, D. P. Byar, D. K. Corle, S. B. Green, C. Schairer, and J. J. Mulvihill. 1989. Projecting individualized probabilities of developing breast cancer for white females who are being examined annually. *Journal of the National Cancer Institute* 81(24):1879– 1886.

Gillis, P., and S. R. Hursh. 1999. Using behavior moderators to influence CGF command entity effectiveness and performance. *In: Proceedings of the Eighth Conference on Computer Generated Forces and Behavior Representation*, Washington, D.C.: U.S. Department of Defense.

Ginnett, R. C. 1993. Crews as groups; their formation and their leadership. *In: Cockpit Resource Management.* E. L. Wiener, B. G. Kanki and R. L. Helmreich, eds.

Goldin, D. S. 1999a. Letter to Kenneth I. Shine, President, Institute of Medicine. January 5.

Goldin, D. S. 1999b. The cornucopia of the future. Remarks delivered to the Board on

Science, Technology and Economic Policy, National Academy of Sciences, Washington, D.C., October 25.

Gossot, D., G. Buess, A. Cuschieri, E. Leporte, M. Lirici, R. Marvik, D. Meijer, A. Melzer, and M. O. Schurr. 1999. Ultrasonic dissection for endoscopic surgery. *Surgical Endoscopy* 13:412–417.

Gottschalk, L., and G. Gleser. 1969. *The Measurement of Psychological States Through the Content Analysis of Verbal Behavior*. Berkeley: University of California Press.

Gould, R. L. 1989. *Therapeutic Learning Program (Version 5.0).* Santa Monica, Calif.:. Interactive Health Systems.

Greenfield, L., ed. 1996. *Essentials of Surgery: Scientific Principles and Practice.* Philadelphia: Lippincott Williams & Wilkins.

Greenleaf, J. E., R. Bulbulian, E. M.Bernauer, W. L. Haskell, and T. Moore. 1989. Exercise training protocols for astronauts in microgravity. *Journal of Applied Physiology* 67(6):2191–2204.

Guerin, B. 1994. *Analyzing Social Behavior: Behavior Analysis and the Social Sciences.* Reno, Nev.: Context Press.

Gundel, A., V. Nalishiti, E. Reucher, M. Vejvoda, and J. Zulley. 1993. Sleep and circadian rhythm during a short space mission. *Clinical Investigation* 71:718–724.

Gunderson, E. K. E. 1966a. *Selection for Antarctic Service.* Unit Report No. 66-15. San Diego, Calif.: Medical Neuropsychiatric Research Unit, U.S. Navy.

Gunderson, E. K. E. 1966b. *Adaptation to Extreme Environments: Prediction of Performance.* Unit Report No. 66-17. San Diego, Calif.: Medical Neuropsychiatric Research Unit, U.S. Navy.

Gunderson, E. K. E., ed. 1974a. *Human Adaptability to Antarctic Conditions.* Antarctic Research Series, Vol. 22. Washington, D.C.: American Geophysical Union.

Gunderson, E. K. E. 1974b. Introduction. *In: Human Adaptability to Antarctic Conditions.* E. K. E. Gunderson, ed. Washington, D.C.: American Geophysical Union.

Hackman, J. 1990. *Groups That Work (and Those That Don't).* San Francisco: Jossey-Bass.

Hackman, J. 1998. Why teams don't work. *In: Theory and Research in Small Groups.* R. Tindale, ed. New York: Plenum.

Hackman, J. R. 1993. Teams, leaders, and organizations: new directions for crew-oriented flight training, pp. 47–69. *In: Cockpit Resource Management.* E. L. Wiener, B. G. Kanki, and R. L. Helmreich, eds. San Diego, Calif.: Academic Press.

Hackman, J. R., and R. E. Walton. 1986. Learning groups in organizations, pp. 72–119. *In: Designing Effective Work Groups.* S. Goodman and Associates, eds. San Francisco: Jossey-Bass.

Hargens, A. R. 1994. Recent bed rest results and countermeasure development at NASA. *Acta Physiologica Scandinavica Supplementum* 616:103–114.

Harlow, H. J., T. Lohuis, T. D. I. Beck, and P. A. Iaizzo. 2001. Muscle strength in overwintering bears: unlike humans, bears retain their muscle tone when moribund for long periods. *Nature* 409(6823):997.

Harris, B. A., R. D. Billica, S. L. Bishop, T. Blackwell, C. S. Layne, D. L. Harm, G. R. Sandoz, and E. C. Rosenow. 1997. Physical examination during space flight. *Mayo Clinic Proceedings* 72(4):301–308.

Harris, J. R. 1991. *Breast Diseases.* Philadelphia: Lippincott.

Harris, J. R., M. E. Lippman, U. Veronesi, and W. Willett. 1992. Breast cancer. *New England Journal of Medicine* 327(5):319–328.

Harrison, A. A., Y. A. Clearwater, and C. P. McKay, eds. 1991. *From Antarctica to Outer Space: Life in Isolation and Confinement.* New York: Springer-Verlag.

Hartley, A. A. 1999. Attention. *In: The Handbook of Aging and Cognition.* F. I. M. Craik and T. A. Salthouse, eds. Hillsdale, N.J.: Erlbaum.

Heifetz, R. A. 1998. *Leadership Without Easy Answers.* Cambridge, Mass.: Harvard University Press.

Heissler, E., F. S. Fischer, S. Bolouri, T. Lehman, W. Mathar, A. Gebhardt, W. Lanksch, and J. Bier. 1998. Custom-made cast titanium implants produced with CAD/CAM for the reconstruction of cranium defects. *International Journal of Oral and Maxillofacial Surgery* 27:334–338.

Helmreich, R. L. 1973. Psychological research in Tektite 2. *Man Environment System* 3:125–127.

Hoffman, S. J., and D. I. Kaplan, eds. 1997. *Human Exploration of Mars: The Reference Mission of the NASA Mars Exploration Study Team.* NASA Special Publication 6107. Houston: National Aeronautics and Space Administration.

Hogan, J., S. L. Rybicki, S. J. Motowidlo, and W. C. Borman. 1998. Relations between contextual performance, personality, and occupational advancement. *Human Performance* 11:189–207.

Holick, M. F. 2000. Microgravity-induced bone loss—will it limit human space exploration? *The Lancet* 355(9215):1569–1570.

Holland, A. 1997. NASA Operational Program. Presentation to the National Research Council Panel on Human Behavior, Committee on Space Biology and Medicine, May 1, National Academy of Sciences, Washington, D.C.

Holland, A. 2000. Space Flight and Analogue Experience. Presentation to the Institute of Medicine Committee on Creating a Vision for Space Medicine During Travel Beyond Earth Orbit, February 23, Johnson Space Center, Houston.

Holland, A. W., L. Looper, and L. Marcondes-North. 1993. Multicultural Factors in the Space Environment: results of an international shuttle crew debrief. *Aviation, Space and Environmental Medicine* 64:196–200.

Hunter, J. B. 1999. Food system challenges for long-duration space missions. *In*: *Proceedings of the First Biennial Space Biomedical Investigators' Workshop*, January 11–13. Houston: National Aeronautics and Space Administration.

ICRP (International Commission on Radiological Protection). 1969. *Radiosensitivity and Spatial Distribution of Dose.* New York: Pergamon.

IOM (Institute of Medicine). 1994. *Careers in Clinical Research.* Washington, D.C.: National Academy Press.

IOM. 1998. *Statement on Quality of Care: National Roundtable on Health Care Quality—The Urgent Need to Improve Health Care Quality.* Washington, D.C.: National Academy Press.

IOM. 1999a. *Fluid Resuscitation: State of the Science for Treating Combat Casualties and Civilian Injuries.* A. Pope, G. French, and D. E. Longnecker, eds. Washington, D.C.: National Academy Press.

IOM. 1999b. *Measuring the Quality of Health Care.* Washington, D.C.: National Academy Press.

IOM. 2000. *To Err is Human: Building a Safer Health System*. L. T. Kohn, J. M. Corrigan, and M. S. Donaldson, eds. Washington, D.C.: National Academy Press.
IOM. 2001a. *Crossing the Quality Chasm: A New Health System for the 21st Century*. Washington, D.C.: National Academy Press.
IOM. 2001b. *Envisioning the National Health Care Quality Report*. M. P. Hurtado, E. K. Swift, and J. M. Corrigan, eds. Washington, D.C.: National Academy Press.
IOM. 2001c. *Exploring the Biological Contributions to Human Health: Does Sex Matter?* Washington, D.C.: National Academy Press.
IOM. 2001d. *Multiple Sclerosis: Current Status and Strategies for the Future*. Washington, D.C.: National Academy Press.
IOM. 2001e. *Small Clinical Trials: Issues and Challenges*. C. H. Evans, Jr., and S. T. Ildstad, eds. Washington, D.C.: National Academy Press.
James, J. T., T. F. Limero, H. J. Leano, J. F. Boyd, and P. A. Covington. 1994. Volatile organic contaminants found in the habitable environment of the space shuttle: STS-26 to STS-55. *Aviation, Space and Environmental Medicine* 65(9):851–857.
Jennings, R. T., and J. P. Bagian. 1996. Musculoskeletal injury review in the US space program. *Aviation, Space and Environmental Medicine* 67(8):762–766.
Jennings, R. T., and E. S. Baker. 2000. Gynecological and reproductive issues for women in space: a review. *Obstetrical and Gynecological Survey* 55(2):109–116.
Jennings, R. T., and P. A. Santy. 1990. Reproduction in space environment. Part II. Concerns for human reproduction. *Obstetrical and Gynecological Survey* 45(1):7–17.
Johnson, P. C., T. B. Driscoll, and C. L. Fisher. 1977. Blood volume changes. *In: Biomedical Results from Skylab*. NASA Special Publication SP-377. Washington, D.C.: National Aeronautics and Space Administration.
Johnson Space Center Institutional Review Board. 1996. *Guidelines for Investigators Proposing Human Research for Space Flight and Related Investigations*. NASA Document JSC 20483, Revision B. Houston: Johnson Space Center, National Aeronautics and Space Administration.
Johnston, R. S., and L. F. Dietlein, eds. 1977. *Biomedical Results from Skylab*. NASA Special Publication SP-377. Washington, D.C.: National Aeronautics and Space Administration.
Judge, T. A., and J. E. Bono. 2000. Five-factor model of personality and transformational leadership. *Journal of Applied Psychology* 85:751–765.
Kabanoff, B. 1980. Work and non-work: a review of models, methods, and findings. *Psychological Bulletin* 88:60–77.
Kanas, N. 1991. Psychological support for cosmonauts. *Aviation, Space and Environmental Medicine* 62:353–355.
Kane, R. L., and G. G. Kay. 1992. Computerized assessment in neuropsychology: a review of tests and test batteries. *Neuropsychology Review* 3:1–17.
Keller, C., J. Brimacombe, M. Giampalmo, A. Kleinsasser, A. Loeckinger, and G. Giampalmo. 2000. Airway management during spaceflight: a comparison of four airway devices in simulated microgravity. *Anesthesiology* 92(5):1237–1241.
Keller, T. 1999. Images of the familiar: individual differences and implicit leadership theories. *Leadership Quarterly* 10(4):589–607.
Kelley, A. D., and N. Kanas. 1992. Crewmember communication in space: a survey of

astronauts and cosmonauts. *Aviation, Space and Environmental Medicine* 63:721–726.

Kelley, A. D., and N. Kanas. 1994. Leisure time activities in space: a survey of astronauts and cosmonauts. *Acta Astronautica* 6:451–457.

Kelly, T., R. Taylor, S. Heishman, and D. Crouch. 1998. Performance measures of behavioral impairment in applied settings. *In: Handbook on Drug Abuse.* S. B. Karch, ed. Boca Raton, Fla.: CRC Press.

Kennaway, D. J., and C. F. Van Dorp. 1991. Free-running rhythms of melatonin, cortisol, electrolytes and sleep in humans in Antarctica. *American Journal of Physiology* 260:R1137– R1144.

Kimzey, S. L., C. L. Fischer, P. C. Johnson, S. E. Ritzmann, and C. E. Mengel. 1975. Hematology and immunology studies. *In: Biomedical Results of Apollo.* R. S. Johnston, L. F. Dietlein, and C. A. Berry, eds. NASA Special Publication 368. Washington, D.C.: National Aeronautics and Space Administration.

Kimzey, S. L., and P. C. Johnson. 1977. Hematological and immunological studies. *In: The Apollo-Soyuz Test Project Medical Report.* A. E. Nicogossian, ed. NASA Special Publication 411. Washington, D.C.: National Aeronautics and Space Administration.

Kirkpatrick, A. W., M. R. Campbell, O. L. Novinkov, I. B. Goncharov, and I. V. Kovachevich. 1997. Blunt trauma and operative care in microgravity: a review of microgravity physiology and surgical investigations with implications for critical care and operative treatment in space. *Journal of the American College of Surgeons* 184(5):441–453.

Kirsch, K. A., L. Rocker, O. H. Gauer, R. Krause, C. Leach, H. J. Wicke, and R. Landry. 1984. Venous pressure in man during weightlessness. *Science* 225(4658):218–219.

Korolev, I. N., and N. Z. Zagorskaia. 1996. The effect of infrared laser radiation of different frequencies on the healing of skin wounds. *Vopr Kurortol Fizioter Lech Kult* 3:8–10.

Koros, A. S. 1991a. *Crewmember Evaluation of Vibration on STS-40/SLS-1.* Houston: National Aeronautics and Space Administration.

Koros, A. S. 1991b. *Objective and Subjective Measures of Noise on STS-40/SLS-1.* Houston: National Aeronautics and Space Administration.

Koros, A. S., S. C. Adam, and C. D. Wheelwright. 1993. A human factors evaluation: noise and its effects on shuttle crewmembers during STS-50/USML-1. NASA Technical Memorandum 104775. Houston: National Aeronautics and Space Administration.

Koslovskaya, I. B., V. A. Barmin, V. I. Stepantsov, and N. M. Kharitinov. 1990. Results of studies of motor functions in long-term spaceflights. *Physiologist* 33:S1–S3.

Kosslyn, S. M., and O. Koenig. 1995. *Wet Mind: The New Cognitive Neuroscience.* New York: Free Press.

Koury, M. J., and M. C. Bondurant. 1990. Erythropoietin retards DNA breakdown and prevents programmed cell death in erythroid progenitor cells. *Science* 248:378–381.

Kraemer, K., and J. King, 1988. Computer-based systems for cooperative work and group decision making. *ACM Computing Surveys* 20:115–146.

Lane, H. W., and D. A. Schoeller. 2000. *Nutrition in Spaceflight and Weightlessness Models.* Washington, D.C.: CRC Press.

Lane, H. W., P. A. Whitson, L. Putcha, E. Baker, S. M. Smith, K. Stewart, R. J. Gretebeck,

R. R. Nimmagudda, D. A. Schoeller, J. Davis-Street, R. A. Pietrzyk, D. E. DeKerlegand, C. Y. C. Pak, and D. W. A. Bourne. 1999. Regulatory physiology. *In: Extended Duration Orbiter Medical Project.* Final Report 1989–1995. C. F. Sawin, G. R. Taylor, and W. L. Smith, eds. NASA Special Publication SP-1999-534. Houston: National Aeronautics and Space Administration.

Lapchine, L., N. Moatti, G. Gassett, G. Richoilley, J. Templier, and R. Tixador. 1986. Antibiotic activity in space. *Drugs Under Experimental and Clinical Research* 12(12):933–938.

Lattal, K. A., and M. Perone. 1998. *Handbook of Research Methods in Human Operant Behavior.* Englewood Cliffs, N.J.: Prentice-Hall.

Lauzon, A. M., A. R. Elliott, M. Paiva, J. B. West, and G. K. Prisk. 1998. Cardiogenic oscillation phase relationships during single-breath tests performed in microgravity. *Journal of Applied Physiology* 84(2):661–668.

Lazarus, A. A. 2000. Multimodal replenishment. *Professional Psychology: Research and Practice* 31(1):93–94

Leach, C. S., and P. C. Johnson. 1984. Influence of spaceflight on erythrokinetics in man. *Science* 225:216–218.

Leach, C. S., C. P. Alfrey, W. N. Suki, J. I. Leonard, P. C. Rambaut, D. Inners, S. M. Smith, H. W. Lane, and J. M. Krauhs. 1996. Regulation of body fluid compartments during short-term space flight. *Journal of Applied Physiology* 81(1):105–116.

Lebedev, V. 1988. *Diary of a Cosmonaut: 211 Days in Space.* College Station, Tex.: Phytoresource Research Information Service.

LeBlanc, A., V. Schneider, L. Shackelford, S. West, V. Ogavov, A. Bakulin, and L. Veronin. 1996. Bone mineral and lean tissue loss after long duration spaceflight. *Journal of Bone and Mineral Research* 11(Suppl. 1):S323.

Lemaire, J. F., D. Heynderick, and D. N. Baker, eds. 1996. *Radiation Belt Models: Models and Standards.* Geophysical Monograph 97. Washington, D.C.: American Geophysical Union.

Letaw, J. R., R. Silberberg, and C. H. Tsao. 1987. Radiation hazards on space missions. *Nature* 330(6150):709–710.

Letaw, J. R., R. Silberberg, and C. H. Tsao. 1988. Galactic cosmic radiation doses to astronauts outside the magnetosphere. *In: Terrestrial Space Radiation and Its Biological Effects.* P. D. McCormack, C. E. Swenberg, and H. Bucker, eds. New York: Plenum.

Levine, B. D., L. D. Lane, D. E. Watenpaugh, F. A. Gaffney, J. C. Buckey, and C. G. Blomqvist. 1996. Maximal exercise performance after adaptation to microgravity. *Journal of Applied Physiology* 81(2):686–694.

Levine, R. J. 1986. *Ethics and Regulation of Clinical Research.* New Haven: Yale University Press.

Levine, R. J. 1991. Informed consent: some challenges to the universal validity of the Western model. *Law, Medicine and Health Care* 19(3–4):207–213.

Lewin, R., and B. Regine. 2000. *The Soul at Work: Embracing Complexity Science for Business Success.* New York: Simon and Schuster.

Lieberman, P., A. Protopappas, and B. Kanki. 1995. Speech production and cognitive deficits on Mt. Everest. *Aviation, Space and Environmental Medicine* 66:857–864.

Linenger, J. M. 2000. *Off the Planet.* New York: McGraw-Hill.

Lipsey, M. W. 1993. The efficacy of psychological, educational and behavioral treatment confirmation from meta-analysis. *American Psychologist* 48(12):1181–1209.

Liu, H. 1996. Pulsed electromagnetic fields influence hyaline cartilage extracellular matrix composition without affecting molecular structure. *Osteoarthritis and Cartilage* 4(1):63–76.

Lovejoy, J. C., S. R. Smith, J. J. Zachwieja, G. A. Bray, M. M. Windhauser, P. J. Wickersham, J. D. Veldhuis, R. Tulley, and J. A. De La Bretonne. 1999. Low-dose T_3 improves the bed rest model of simulated weightlessness in men and women. *American Journal of Physiology* 277(2):E370–E379.

Luchette, F. A., A. P. Borzotta, M. A. Croce, P. A. O'Neill, D. H. Whittmann, C. D. Mullins, F. Palumbo, and M. D. Pasquale. 2000. Practice management guidelines for prophylactic antibiotic use in penetrating abdominal trauma: the EAST Practice Management Guidelines Work Group. *Journal of Trauma—Injury Infection and Critical Care* 48:508–518.

Lugg, D. J. 1979. *Appendicitis in Polar Regions.* Diploma in Polar Studies Thesis, University of Cambridge, Cambridge, United Kingdom.

Lugg, D. J. 2000. Antarctic medicine. *Journal of the American Medical Association* 283(16):2082–2084.

Lugg, D. J., and M. Shepanek. 1999. Space analogue studies in Antarctica. *Acta Astronautica* 44(7):693–699.

Mandel, I. D. 1996. Caries prevention: current strategies, new directions. *Journal of the American Dental Association* 127:1477–1488.

Mandel, I. D. 2000. Dental Caries: Initiation, Progression and Prevention. Presentation to the Institute of Medicine Committee on Creating a Vision for Space Medicine During Travel Beyond Earth Orbit, April 17, National Academy of Sciences, Washington, D.C.

Maniscalco-Theberge, M. E., and D. C. Elliott. 1999. Virtual reality, robotics, and other wizardy in 21st century trauma care. *Surgical Clinics of North America* 79:1241–1248.

Manzey, D., and B. Lorenz. 1998. Performance during short-term and long-term spaceflight. *Brain Research Reviews* 28(1):215–221.

Manzey, D., T. B. Lorenz, H. Heuers, and J. Sangals. 2000. Impairments of manual tracking performance during spaceflight: more converging evidence from a 20-day space mission. *Ergonomics* 43(5):589–609.

Markham, C. H., and S. G. Diamond. 1999. Otolith-ocular torsion is modified in novel g states. *In*: *Proceedings of the First Biennial Space Biomedical Investigators' Workshop*, January 11–13. Houston: National Aeronautics and Space Administration.

Markowitz, S. 1992. The role of surveillance in occupational health. *In: Environmental and Occupational Medicine,* 2nd ed. W. N. Rom, ed. Boston: Little Brown.

Marshburn, T. 2000a. Pathophysiology of Long-Duration Spaceflight. Presentation to the Institute of Medicine Committee on Creating a Vision for Space Medicine During Travel Beyond Earth Orbit, February 22, Johnson Space Center, Houston.

Marshburn, T. 2000b. Phase 1/*Mir* Clinical Experience. Presentation to the Institute of Medicine Committee on Creating a Vision for Space Medicine During Travel Beyond Earth Orbit, February 22, Johnson Space Center, Houston.

Maruyama, M. 1978. Settlements in space. *In: Handbook of Future Research.* J. Fowles, ed. Westport, Conn.: Greenwood Press.

McCarter, F. D., F. A. Luchette, M. Molloy, J. M. Hurst, K. Davis, Jr., J. A. Johannigman, S. B. Frame, and J. E. Fischer. 2000. Institutional and individual learning curves for focused abdominal ultrasound for trauma: cumulative sum analysis. *Annals of Surgery* 231:689–700.

McCuaig, K. 1992. Aseptic technique in microgravity. *Surgery, Gynecology and Obstetrics* 175(5):466–476.

McGrath, J. 1984. *Groups: Interaction and Performance.* Englewood Cliffs, N.J.: Prentice-Hall.

McGrath, J. 1997. Small group research, that once and future field: an interpretation of the past with an eye to the future. *Group Dynamics* 1(1):7–27.

McGrath, J., and I. Altman. 1966. *Small Group Research: A Synthesis and Critique of the Field.* New York: Holt, Rinehart and Winston.

Mehta, S. K., S. K. Mishra, and D. L. Pierson. 1996. Evaluation of three portable samplers for monitoring airborne fungi. *Applied and Environmental Microbiology* 62(5):1835–1838.

Meleshko, G. I., Y. Y. Shepelev, M. M. Averner, and T. Volk. 1994. Biological life support systems, p. 383. *In: Space Biology and Medicine*, Vol. II. *Life Support and Habitability.* A. E. Nicogossian, S. R. Mohler, O. G. Gazenko, and A. I. Grigoryev, eds. Washington, D.C.: American Institute of Aeronautics and Astronautics, and Moscow: Nauka Press.

Menasche, A., and R. J. Levine. 1995. FDA revises informed consent regulations for emergency research. *IRB: A Review of Human Subjects Research* 17(5–6):19–22.

Mikhailov, V. M., Y. D. Pometov, and V. A. Andretsov. 1984. LBNP training of crew members of the Saliut-6 orbital station. *Kosmicheskaia Biologiia Aviakosmicheskaia Meditsina* 18(6):29–33.

Miller, F. J., D. T. Ahola, P. A. Bretzman, and D. J. Fillmore. 1997. Percutaneous management of hepatic abscess: a perspective by interventional radiologists. *Journal of Vascular Intervention and Radiology* 8(2):241–247.

Monk, T. H., D. J. Buysse, B. D. Billy, K. S. Kennedy, and L. M. Willrich. 1998. Sleep and circadian rhythms in four orbiting astronauts. *Journal of Biological Rhythms* 13:188–201.

Montgomery, R. S., and S. E. Wilson. 1996. Intraabdominal abscesses: image-guided diagnosis and therapy. *Clinical Infectious Diseases* 23(1):28–36.

Mount, F. E., and T. Foley. 1999. Assessment of human factors. *In: Extended Duration Orbiter Medical Project.* Final Report 1989–1995. C. F. Sawin, G. R. Taylor, and W. L. Smith, eds. NASA Special Publication SP-1999-534. Houston: National Aeronautics and Space Administration.

Mount, M. K., M. R. Barrisck, and G. L. Stewart. 1998. Five-factor model of personality and performance in jobs involving interpersonal interactions. *Human Performance* 11:145–165.

Mullington, J., D. Hermann, F. Holsboer, and T. Pollmacher. 1996. Age-dependent suppression of nocturnal growth hormone levels during sleep deprivation. *Neuroendocrinology* 64(3):233–241.

Mulvihill, K. 1999. Zipping up: a substitute for stitches and staples. *The New York Times*, December 28, page 6.

Mundt, C. 1999. Biotelemetry. *In: Proceedings of the First Biennial Space Biomedical*

Investigators' Workshop, January 11–13. Houston: National Aeronautics and Space Administration.

Murphy, L. R. 1996. Stress management in work settings: a critical review of the health effects. *American Journal of Health Promotion* 11(2):112–135.

Nakamoto, D. A., and J. R. Haaga. 1995. Percutaneous drainage of postoperative and intra-abdominal abscesses and collections. *In: Current Techniques in Interventional Radiology*. C. Cope, ed. Philadelphia: Current Medicine.

NASA (National Aeronautics and Space Administration). 1990. *Mission Operation Directorate.* Houston: National Aeronautics and Space Administration.

NASA. 1997. *Task Force on Countermeasures: Final Report.* Washington, D.C.: National Aeronautics and Space Administration.

NASA. 1998a. *Astronaut Medical Evaluation Requirements Document*, Appendix B. Houston: National Aeronautics and Space Administration.

NASA. 1998b. *Clinical Care Capability Development Project.* JSC No. 28358. Houston: National Aeronautics and Space Administration.

NASA. 1999. *Medical Informatics and Telemedicine for Space Flight Workshop Report*, November 9–10. Washington, D.C.: National Aeronautics and Space Administration.

NASA. 2000a. *Astronaut Physical Training and Rehabilitation Research Workshop Report*, September 14–15. Houston: National Aeronautics and Space Administration.

NASA. 2000b. Medical care during spaceflight: Views of physician astronauts. Presentation to the Institute of Medicine Committee on Creating a Vision for Space Medicine During Travel Beyond Earth Orbit, July 27, Woods Hole, MA.

Nelson, P. D. 1964a. Similarities and differences among leaders and followers. *Journal of Applied Psychology* 63:161–167.

Nelson, P. D. 1964b. Supervisor esteem and personnel evaluation. *Applied Psychology* 48:106–109.

Nelson, P. D. 1965. *Psychological Aspects of Antarctic Living.* Unit Report No. 64-28. San Diego, Calif.: Medical Neuropsychiatric Research Unit, U.S. Navy.

Nelson, P. D. 1973. Indirect observation in groups. *In: Man in Isolation and Confinement.* J. E. Rasnussen ed. Chicago: Aldine Publishing Co.

Newman, M. G. 1997. Comparison of palmtop, computer-assisted brief cognitive behavioral treatment to cognitive behavioral treatment for panic disorder. *Journal of Consulting and Clinical Psychology* 65(12):178–183.

Nicholas, J. 1987. Small groups in orbit: group interaction and crew performance on space station. *Aviation, Space and Environmental Medicine* 58:1009–1013.

Nicholas, J. M., and L. W. Penwell 1995. A proposed profile of the effective leader in human spaceflight based on findings from analog environments. *Aviation, Space and Environmental Medicine* 66(1):63–72.

Nicogossian, A. E., and J. F. Parker, Jr. 1982. *Space Physiology and Medicine.* NASA Special Publication SP-447. Washington, D.C.: National Aeronautics and Space Administration.

Nicogossian, A. E., C. L. Huntoon, and S. L. Pool. 1994. *Space Physiology and Medicine.* Fourth Edtion. Malvern, Pa.: Lea & Febiger.

Nicogossian, A. E., C. L. Huntoon, and S. L. Pool. *Space Physiology and Medicine.* Fifth Edtion. In press.

Nightingale, S. L. 1996. From the Food and Drug Administration: exception from informed consent requirements for emergency research. *Journal of the American Medical Association* 276(20):1632.

Noy, S. 1987. Combat psychiatry: the American and Israeli experience, pp. 69–86. *In: Contemporary Studies in Combat Psychiatry.* G. Belenky, ed. Westort: Greenwood Press.

NRC (National Research Council). 1992. *Guidelines for Developing Spacecraft Maximum Allowable Concentrations of Space Station Contaminants.* Washington, D.C.: National Academy Press.

NRC. 2000. *Methods for Developing Spacecraft Water Exposure Guidelines.* Washington, D.C.: National Academy Press.

Office of Strategic Services. 1948. *Assessment of Men.* New York: Rinehart.

O'Kane, R. H. 1977. *Clear the Decks: The War Patrols of the U.S.S. Tang.* Chicago: Rand McNally.

Okumura, H., L.-H. Chen, Y. Yokoe, S. Tsutsumi, and M. Oka. 1999. CAD/CAM fabrication of occlusal splints for orthognatic surgery. *Journal of Clinical Orthodontics* 33(4):231–235.

Oldham, J. A., T. E. Howe, T. Patterson, G. P. Smith, and R. C. Tallis. 1995. Electrotherapeutic rehabilitation of the quadriceps in elderly osteoarthritic patients: a double blind assessment of patterned neuromuscular stimulation. *Clinical Rehabilitation* 9(1):10–20.

Palinkas, L. A. 1990. Psychosocial effects of adjustment in Antarctica: lessons for long-duration spaceflight. *Journal of Spacecraft Rockets* 27:471–477.

Palinkas, L. A. 1992. Going to extremes: the cultural context of stress, illness and coping in Antarctica. *Social Science Medicine* 35:651–664.

Palinkas, L. 2000a. Behavior and performance on long-duration spaceflights: evidence from analogue environments. *Aviation, Space and Environmental Medicine,* 71:1–8.

Palinkas, L. A. 2000b. Summary of research issues in behavior and performance in isolated and confined extreme environments. *Aviation, Space and Environmental Medicine.* 71(9 Suppl.):A48–A50.

Palinkas, L. A., P. Suedfeld, and G. D. Steel. 1995. Psychological functioning among members of a small polar expedition. *Aviation, Space and Environmental Medicine* 50:1591–1596.

Palinkas, L. A., J. C. Johnson, J. S. Boster, and M. Houseal. 1998. Longitudinal studies of behavior and performance during a winter at the South Pole. *Aviation, Space and Environmental Medicine* 69(1):73–77.

Palinkas, L. A., E. K. Gunderson, A. W. Holland, C. Miller, and J. C. Johnson. 2000. Predictors of behavior and performance in extreme environments: the Antarctic space analogue program. *Aviation, Space and Environmental Medicine.* 71(6):619–625.

Paloski, W.H. 2000. Countermeasure evaluation and validation project. . Presentation to the Institute of Medicine Committee on Creating a Vision for Space Medicine During Travel Beyond Earth Orbit, February 22, Johnson Space Center, Houston.

Pantalos, G. M., M. K. Sharp, S. J. Woodruff, D. S. O'Leary, R. Lorange, S. D. Everett, T. E. Bennett, and T. Shurfranz. 1998. Influence of gravity on cardiac performance. *Annals of Biomedical Engineering* 26(6):931–943.

Pardoe, R. 1965. A ruptured intracranial aneurysm in Antarctica. *The Medical Journal of Australia* 1:344–350.

Parmet, J. L., P. Colonna-Romano, J. C. Harrow, F. Miller, J. Gonzales, and H. Rosenberg. 1998. The laryngeal mask airway reliably provides rescue ventilation in cases of unanticipated difficult tracheal intubation along with difficult mask ventilation. *Anesthesia and Analgesia* 87(3):661–665.

Payne, D. A., S. K. Mehta, S. K. Trying, R. P. Stowe, and D. L. Pierson. 1999. Incidence of Epstein-Barr virus in astronaut saliva during space flight. *Aviation, Space and Environmental Medicine* 70(12):1211–1213.

Peterson, L. E., L. J. Pepper, P. B. Hamm, and S. L. Gilbert. 1993. Longitudinal study of astronaut health: mortality in the years 1959–1991. *Radiation Research* 133:257–264.

Pickett, E., E. Kuniholm, A. Protopappas, J. Friedman, and P. Lieberman. 1998. Selective speech motor, syntax and cognitive deficits associated with bilateral damage to the head of the caudate nucleus and the putamen: a single case study. *Neuropsychologia* 36:273–288.

Pierson, D., J. James, D. Russo, T. Limero, S. Beck, and T. Groves. 1999. Environmental health. *In: Extended Duration Orbiter Medical Project.* Final Report 1989–1995. NASA Special Publication SP-1999-534. C. F. Sawin, G. R. Taylor, and W. L. Smith, eds. Houston: National Aeronautics and Space Administration.

Pinsonneault, A., and K. Kraemer. 1990. The effects of electronic meetings on group process and outcomes: an assessment of the empirical research. *European Journal of Operations Research* 46:143–161.

Plakhuta-Plakutina, G. I. 1977. State of spermatogenesis in rats flown aboard the biosatellite Cosmos-690. *Aviation, Space and Environmental Medicine* 48(1):12–15.

Pollitt, J., and K. Flynn. 1999. NASA Ames Research Center R & D Services Directorate Biomedical Systems Development. *In: Proceedings of the First Biennial Space Biomedical Investigators' Workshop,* January 11–13. Houston: National Aeronautics and Space Administration.

Pool, S. 2000. The Longitudinal Study of Astronaut Health. Presentation to the Institute of Medicine Committee on Creating a Vision for Space Medicine During Travel Beyond Earth Orbit, February 22, Johnson Space Center, Houston.

Power, K. G., R. J. Simpson, V. Swanson, L. A. Wallace, A. T. C. Feistner, and D. Sharp. 1990. A controlled comparison of cognitive-behavior therapy, diazepam and placebo, alone and in combination, for treatment of generalized anxiety disorder. *Journal of Anxiety Disorders* 4:267–292.

Pratt, D. R., S. M. Pratt, M. S. Waldrop, P. T. Barham, J. F. Ehlert, and C. A. Chrislip. 1997. Humans in large-scale, real-time networked virtual environments. *Presence* 6(5):547–564.

President's Advisory Commission on Consumer Protection and Quality in the Health Care Industry. 1998. *Quality First: Better Health Care for All Americans.* Washington, D.C.: U.S. Department of Health and Human Services.

Priddy, R. E. 1985. An "acute abdomen" in Antarctica. The problems of diagnosis and management. *The Medical Journal of Australia* 143:108–111.

Prisk, G. K., H. J. Guy, A. R. Elliott, R. A. Deutschman III, and J. B. West. 1993. Pulmonary diffusing capacity, capillary blood volume, and cardiac output during sustained microgravity. *Journal of Applied Physiology* 75(1):15–26.

Prisk, G. K., A. R. Elliott, H. J. B. Guy, J. M. Kosonen, and J. B. West. 1995a. Pulmonary

gas exchange and its determinants during sustained microgravity on Spacelabs SLS-1 and SLS-2. *Journal of Applied Physiology* 79(4):1290–1298.
Prisk, G. K., H. J. B. Guy, A. R. Elliott, M. Paiva, and J. B. West. 1995b. Ventilatory inhomogeneity determined from multiple-breath washouts during sustained microgravity on Spacelab SLS-1. *Journal of Applied Physiology* 78(2):597–607.
Putcha, L., K. L. Berens, T. H. Marshburn, H. J. Ortega, and R. D. Billica. 1999. Pharmaceutical use by U.S. astronauts on space shuttle missions. *Aviation, Space and Environmental Medicine* 70(7):705–708.
Quinn, J., G. Wells, T. Sutcliffe, M. Jarmuske, J. Maw, I. Stiell, and P. Johns. 1998. Tissue adhesive versus suture wound repair at 1 year: randomized clinical trial correlating early, 3-month, and 1-year cosmetic outcome. *Annals of Emergency Medicine* 32:645–649.
Radloff, R.W., and R. Helmreich. 1973. *Groups Under Stress: Psychological Research in Sealab II.* New York: Appleton-Century-Crofts.
Rayman, R. S. 1997. An ethical dilemma. *Aviation, Space and Environmental Medicine* 68(10):973.
Rayman, R. S. 1998. Aerospace medicine and ethics. *Aviation, Space and Environmental Medicine* 69(12):1223.
Reed, H. L., E. D. Silverman, K. M. M. Shakir, R. Dons, K. D. Burman, and J. T. O'Brian. 1990. Change in serum triiodothyronine (T3) kinetics after prolonged Antarctic residence: the polar T3 syndrome. *Journal of Clinical Endocrinology and Metabolism* 70:965–974.
Riley, D. A., G. R. Slocum, J. L. W. Bain, F. R. Sedlak, T. E. Sowa, and J. W. Mellender. 1990. Rat hind limb unloading: soleus histochemistry, ultrastructure, and electromyography. *Journal of Applied Physiology* 69:58–66.
Riley, D. A., S. Ellis, G. R. Slocum, F. R. Sedlak, J. L. W. Bain, B. B. Krippendorf, C. T. Lehman, M. Y. Macias, J. L. Thompson, K. Vijayan, and J. A. DeBruin. 1996. In-flight and postflight changes in skeletal muscles of SLS-1 and SLS-2 spaceflown rats. *Journal of Applied Physiology* 81:133–144.
Rosen, J. B., and J. Schulkin. 1998. From normal fear to pathological anxiety. *Psychological Review* 105:325–350.
Ross, J. K., J. Arendt, J. Horne, and W. Haston. 1995. Night-shift work in Antarctica: sleep characteristics and bright light treatment. *Physiology and Behavior* 57:1169–1174.
Sack, D. 1998. Total Atlantic Fleet Medical Evacuations from Submarines, 1993–1996. Paper presented at Undersea and Hyperbaric Medicine Meeting, Seattle, Wash.
Sagan, C. 1980. *Cosmos*. New York: Random House.
Salgado, J. F. 1998. Big Five personality dimensions and job performance in Army and civil occupations: a European perspective. *Human Performance* 11:271–278.
Salthouse, T. A. 1991. *Theoretical Perspectives on Cognitive Aging.* Hillsdale, N.J.: Erlbaum.
Sampson, J. A. 1927. Peritoneal endometriosis due to the menstrual dissemination of endometrial tissue into the peritoneal cavity. *American Journal of Obstetrics and Gynecology* 14:422.
Sandal, G. M., I. M. Endresen, R. Vaernes, and H. Ursin. 1999. Personality and coping strategies during submarine missions. *Military Psychology* 11(4):381–404.

Sandal, G. M., R. Vaernes, and H. Ursin. 1995. Interpersonal relations during simulated space missions. *Aviation, Space and Environmental Medicine* 66:617–624.

Santy, P. A. 1994. *Choosing the Right Stuff: The Psychological Selection of Astronauts and Cosmonauts*. Westport, Conn.: Praeger.

Santy, P. A., H. Kapanka, J. R. Davis, and D. F. Stewart. 1988. Analysis of sleep on shuttle missions. *Aviation, Space and Environmental Medicine* 59:1094–1097.

Satava, R. M. 1997. Lest we forget the future. *Journal of the American College of Surgeons* 184(5):519–520.

Satava, R. M., and S. B. Jones. 1998. Telepresence surgery. *In*: *Cybersurgery*. R. M. Satava, ed. New York: Wiley-Liss.

Sawin, C. 2000. Countermeasure Evaluation: Extended Duration Orbiter Medical Program. Presentation to the Institute of Medicine Committee on Creating a Vision for Space Medicine During Travel Beyond Earth Orbit, February 23, Johnson Space Center, Houston.

Schneider, V. S., A. D. LeBlanc, and L. C. Taggart. 1994. Bone and mineral metabolism. *In: Space Physiology and Medicine*. A. E. Nicogossian, C. L. Huntoon, and S. L. Pool, eds. Malvern, Pa.: Lea & Febiger.

Schonfeld, J. 1999. Neurolab Biotelemetry System (NBS). *In: Proceedings of the First Biennial Space Biomedical Investigators' Workshop*, January 11–13. Houston: National Aeronautics and Space Administration.

Schwartz, G. E. 1979. Biofeedback and the behavioral treatment of disorders of disregulation. *Yale Journal of Biology and Medicine* 52(6):581–596.

Scott, G. C., J. G. Letourneau, J. M. Berman, and T. R. Beidle. 1997. Drainage of abdominal abscesses. *In: Interventional Radiology*, Vol. 2. W. R. Castaneda-Zuniga, ed. Baltimore: Williams & Wilkins.

Scott, R. B., R. W. TeLinde, and L. R. Wharton. 1953. Further studies on experimental endometriosis. *American Journal of Obstetrics and Gynecology* 66:1082.

Seddon, R., K. Baldwin, J. Hays, A. LeBlanc, E. Baker, B. Brody, R. Jennings, S. Mark, G. Sonnenfeld, and R. Weller. 1999. Gender-Related Issues in Space Flight Research and Health Care. *In: Enabling a Broader Segment of the Population to Explore, Live, and Work in Space in the 21st Century*. National Space Biomedical Research Institute Workshop Report. An Element of NSBRI Project 99-4 (Cooperative Agreement NCC 9-58). Houston: National Aeronautics and Space Administration.

Sells, S., and E. Gunderson. 1972. A social-system approach to long-duration missions. *In: Human Factors in Long-Duration Spaceflight*. D. B. Lindsley, ed. Washington, D.C.: National Academy of Sciences.

Shapiro, D., and G. E. Schwartz. 1972. Biofeedback and visceral learning: clinical applications. *Seminars in Psychiatry* 4(2):171–184.

Shuler, F. W., C. N. Newman, P. B. Angood, J. G. Tucker, and G. W. Lucas. 1996. Nonoperative management for intra-abdominal abscesses. *The American Surgeon* 62(3):218–222.

Sidell, F. R., E. T. Takafuji, and D. R. Franz, eds. 1991. *Medical Aspects of Chemical and Biological Warfare*. Washington, D.C.: Office of the Surgeon General, TMM Publications.

Silverthorne, C. 2001. Leadership effectiveness and personality: a cross cultural evaluation. *Personality & Individual Differences* 30(2):303–309.

Simanonok, K. E., and J. B. Charles. 1994. Space sickness and fluid shifts: a hypothesis. *Journal of Clinical Pharmacology* 34:652–663

Singer, A. J., J. E. Hollander, and J. V. Quinn. 1997. Evaluation and management of traumatic lacerations. *New England Journal of Medicine* 337(16):1142–1148.

Singer, P., and C. Vann. 1975. Extraterrestrial communities—cultural, legal, political and ethical considerations. *In: Cultures Beyond.* M. Maruyama and A. Harkins, eds. New York: Vintage Books.

Smith, S. M., M. E. Wastney, B. V. Morukov, I. L. Larina, L. E. Nyquist, S. A. Abrams, E. N. Taran, C. Y. Shi, J. L. Nillen, J. E. Davis-Street, B. L. Rice, and H. W. Lane. 1999. Calcium metabolism before, during, and after a 3-mo spaceflight: kinetic and biochemical changes. *American Journal of Physiology* 277(1 Pt 2):R1–R10.

Socrates 1952. *Phaedo*, translated by Benjamin Jowett, in the *Dialogues of Plato*, Vol. 7, *Plato. In: Great Books of the Western World.* R. M. Hutchins, editor-in-chief, Encyclopaedia Britannica, Inc., Chicago, pp. 247-248.

SSB (Space Studies Board) and NRC (National Research Council). 1987. *A Strategy for Space Biology and Medical Science for the 1980s and 1990s.* Washington, D.C.: National Academy Press.

SSB and NRC. 1996. *Radiation Hazards to Crews of Interplanetary Missions.* Washington, D.C.: National Academy Press.

SSB and NRC. 1998a. *A Strategy for Research in Space Biology and Medicine in the New Century.* Washington, D.C.: National Academy Press.

SSB and NRC. 1998b. Bone physiology. *In: A Strategy for Research in Space Biology and Medicine in the New Century.* Washington, D.C.: National Academy Press.

SSB and NRC. 1998c. Cardiovascular and pulmonary systems. *In: A Strategy for Research in Space Biology and Medicine in the New Century.* Washington, D.C.: National Academy Press.

SSB and NRC. 1998d. Radiation hazards. *In: A Strategy for Research in Space Biology and Medicine in the New Century.* Washington, D.C.: National Academy Press.

SSB and NRC. 2000a. *Radiation and the International Space Station. Recommendations to Reduce Risk.* Washington, D.C.: National Academy Press.

SSB and NRC. 2000b. *Review of NASA's Biomedical Research Program.* Washington, D.C.: National Academy Press.

Stein, T. P., M. J. Leskiw, M. D. Schluter, M. R. Donaldson, and I. Larina. 1996. Protein kinetics during and after long term space flight on Mir. *American Journal of Physiology* 276(39):1014–1021.

Stein, T. P., M. J. Leskiw, M. D. Schluter, R. W. Hoyt, H. W. Lane, R. E. Gretebeck, and A. D. LeBlanc. 1999. Energy expenditure and balance during space flight on the shuttle: the LMS mission. *American Journal of Physiology* 276(45):R1739–R1748.

Sternberg, R. J. 2000. The holy grail of general intelligence. *Science* 289:399–401.

Stogatz, S. H., R. E. Kronauer, and C. A. Czeisler. 1987. Circadian pacemaker interferes with sleep onset at specific times each day: role in insomnia. *American Journal of Physiology* 253:R172–R178.

Strollo, F. 1999. Hormonal changes in humans during spaceflight. *Advances in Space Biology and Medicine* 7:99–129.

Stuster, J. 1986. *Space Station Habitability Recommendations Based on a Systematic Comparative Analysis of Analogous Conditions.* NASA-CR-3943. Moffett Field, Calif.: NASA Ames Research Center.

Stuster, J. 1996. *Bold Endeavours: Lessons from Polar and Space Exploration.* Annapolis, Md.: Naval Institute Press.

Sullivan, P., and P. J. Gormley. 1999. The Australian National Antarctic Research Expeditions Health Register. *In: Proceedings of the National Centre for Classification in Health, 6th Annual Conference.* H. K. Peasley, ed. Lidcombe: Australia. National Centre for Classification in Health .

Sullivan, P., P. J. Gormly, D. J. Lugg, and D. J. Watts. 1991. The Australian National Antarctic Research Expeditions Health Register: Three years of operation. *In: Circumpolar Health 90.* B. Postl, ed. Winnipeg, Manitoba, Canada: University of Manitoba Press.

Taggar, S., R. Hackett, and S. Saha. 1999. Leadership emergence in autonomous work teams: antecedents and outcomes. *Personnel Psychology* 52(4):899–926.

Tansey, W. A., J. M. Wilson, and K. E. Schaefer. 1979. Analysis of health data from 10 years of Polaris submarine patrols. *Undersea Biomedical Research, Submarine Supplement* 6:S217–S246.

Taylor, A. J. W. 1987. *Antarctic Psychology.* Wellington, New Zealand: Science Publishing Centre.

Taylor, A. J. W. 1991. The research program of the International Biomedical Expedition (IBEA) to the Antarctic and its implication for research in outer space. *In: From Antarctica to Outer Space: Life in Isolation and Confinement.* A. A. Harrison, Y. A. Clearwater, and C. P. McKay, eds. New York: Springer-Verlag.

Taylor, D. M., and P. J. Gormly. 1997. Emergency medicine in Antarctica. *Emergency Medicine* 9:237–245.

Thomas, T. L., T. I. Hooper, M. Camarca, J. Murray, D. Sack, D. Mole, R. T. Spiro, W. G. Horn, and F. C. Garland. 2000. A method for monitoring the health of U.S. Navy submarine crewmembers during periods of isolation. *Aviation, Space and Environmental Medicine* 71(7):699–705.

Thornton, W. E., T. P. Moore, and S. L. Pool. 1987. Fluid shifts in weightlessness. *Aviation, Space and Environmental Medicine* 58(9 Suppl. Pt 2):A86–A90.

Tietze, K. J., and L. Putcha. 1994. Factors affecting drug bioavailability in space. *Journal of Clinical Pharmacology* 34:671–676.

Tigranjan, R. A., H. Haase, N. F. Kalita, V. M. Ivanov, E. A. Pavlova, B. V. Afonin, L. I. Voronin, and G. Jarsumbeck. 1982. Results of endocrinologic studies of the 3rd international crew of the scientific orbital station complex; Soyuz 29-Salyut 6-Soyuz 31 (joint space flight enterprise of the USSR-GDR). 2. Hormones and biologically active substances in blood. *Endokrinologie* 80(1):37–41.

Tingate, T. R., D. J. Lugg, H. K. Muller, R. P. Stowe, and D. L. Pierson. 1997. Antarctic isolation: immune and viral studies. *Immunology and Cell Biology* 75:275–283.

Truog, R. D., W. M. Robinson, A. Randolph, and A. Morris. 1999. Is informed consent always necessary in randomized controlled trials? *New England Journal of Medicine* 340(10):804–807.

Turkel, S. 1995. *Life on the Screen: Identity in the Age of the Internet*, pp. 103–124. New York: Simon and Schuster.

UHSC (University Health System Consortium). 1999. *Fibrin Sealants.* Oak Brook, Ill.: Clinical Practice Advancement Center, University Health System Consortium.

Vico, L., P. Collet, A. Guignandon, M. Lafage-Proust, T. Thomas, and M. Rehailia. 2000.

Effects of long-term microgravity exposure on cancellous and cortical weight-bearing bones of cosmonauts. *The Lancet* 355(9215):1607–1611.

Vinograd, S. P. 1974. *Studies of Social Group Dynamics Under Isolated Conditions. Objective Summary of the Literature as It Relates to Potential Problems of Long Duration Space Flight.* NASA-CR-2496. Washington, D.C.: National Aeronautics and Space Administration.

Walton, K., C. Heffernan, D. Sulica, and L. Benavides. 1997. Changes in gravity influence rat postnatal motor system development: from simulation to spaceflight. *Gravitation and Space Biology Bulletin* 10:111–118.

Weigelt, J. A., and S. Faro. 1998. Antimicrobial therapy for surgical prophylaxis and for intra-abdominal and gynecologic infections. *American Journal of Surgery* 176:1S–3S.

Weizenbaum, J. 1966. A Computer Program for the Study of Natural Language Communication Between Man and Machine. *Communications of the Association of Computing Machinery* 9:36–45.

Weizenbaum, J. 1979. *Computer Power and Human Reason: From Judgement to Calculation.* New York: Freeman.

West, J. B., A. R. Elliott, H. J. B. Guy, and G. K. Prisk. 1997. Pulmonary function in space. *Journal of the American Medical Association* 277(24):1957–1961.

West, O. 2001. Climbing Everest: Risking Death for a View From the Top. *The New York Times*, June 24, p. 8.11

Weybrew, B. 1991. Thirty years of research on submarines. *In: From Antarctica to Outer Space: Life in Isolation and Confinement.* A. A. Harrison, Y. A. Clearwater, and C. P. McKay, eds. New York: Springer-Verlag.

Wheatcroft, M. 1989. Effects of simulated Skylab missions on the oral health of astronauts. *The Journal of the Greater Houston Dental Society* 61(5):7.

Widrick, J. J., and R. H. Fitts. 1997. Peak force and maximal shortening velocity of soleus fibers after non-weight-bearing and resistance exercise. *Journal of Applied Physiology* 82(1):189–195.

Wilken, D. D. 1969. *Significant Medical Experiences Aboard Polaris Submarines: A Review of 360 Patrols During the Period 1963–1967.* Report No. 560. Washington, D.C.: Naval Submarine Medical Research Library.

Williams, E. 1974. Behavioral technology and behavioral ecology. *Journal of Applied Behavior Analysis* 7:151–165.

Williams, R. 2000. Multinational Medical Policy Board. Presentation to the Institute of Medicine Committee on Creating a Vision for Space Medicine During Travel Beyond Earth Orbit, October 26, National Academy of Sciences, Washington, D.C.

Wirth, O., P. Chase, and K. Munson. 2000. Experimental analysis of human vocal behavior: applications of speech recognition technology. *Journal of the Experimental Analysis of Behavior* 74:363–375.

Wolpe, J. 1958. *Psychotherapy by Reciprocal Inhibition.* Stanford, Calif.: Stanford University Press.

Wood, J. A., D. J. Lugg, D. J. Eksuzian, S. J. Hysong, and D. L. Harm. 1999. Psychological changes in 100-day remote Antarctic field groups. *Environment and Behavior* 31:299–337.

Wright, K., M. E. Jewett, E. Klerman, R. Kronauer, and C. Czeisler. 1999. Circadian entrainment, sleep-wake regulation and neurobehavioral performance under the simulated lighting conditions of long-duration space missions. *In: Proceedings of the*

First Biennial Space Biomedical Investigators' Workshop, January 11–13. Houston: National Aeronautics and Space Administration.

Wroblicka, T. J., and E. Kuligowska. 1998. One-step needle aspiration and lavage for the treatment of abdominal and pelvic abscesses. *American Journal of Roentgenology* 170(5):1197–1203.

Yost, B. 1999. Life Sciences Division Spaceflight Hardware. *In: Proceedings of the First Biennial Space Biomedical Investigators' Workshop*, January 11–13. Houston: National Aeronautics and Space Administration.

Zimmerman, B. J., C. Lindberg, and P. E. Plsek. 1998. *Edgware: Complexity Resources for Healthcare Leaders*. Dallas: VHA Publications.

APPENDIXES

A
Background and Methodology

The committee was asked to assess the evidence base in space-related clinical research and astronaut health care (see the letter from the NASA administrator at the end of this appendix). It began by examining traditional sources of data and information, including research reports in peer-reviewed journals, conference proceedings, prior National Research Council reports, and National Aeronautics and Space Administration (NASA) publications. The committee also consulted with numerous experts in areas that included medical informatics, bone metabolism, dental disease, and the effects on health and behavior of living and working for long periods of time under extreme and remote conditions.

The single greatest source of data and information was NASA. Agency officials briefed committee members during the committee's first meeting in Washington, D.C., in October 1999. They were Arnauld E. Nicogossian, associate director for Life and Microgravity Sciences; David Williams, director, Life Sciences Directorate; and Richard Williams, deputy chief medical officer. Then, in February 2000, committee members visited NASA's Johnson Space Center (JSC) in Houston, Texas, for an additional 2 days of briefings by NASA staff.

Other highlights of the committee's information-gathering efforts are presented in the pages that follow. Altogether, the committee held six public meetings during which it collected information and interacted with invited experts and interested members of the public. Two of these meetings included workshops during which invited experts made presentations on topics of special interest. The committee gratefully acknowledges the interest and generous assistance of all who aided in its information-gathering efforts.

COMMITTEE SITE VISIT TO NASA'S JOHNSON SPACE CENTER

From February 22 to 24, 2000, committee members visited JSC in Houston, Texas, for intensive briefings by NASA administrators, scientists, and clinicians. The JSC faculty made 15 presentations to the full Institute of Medicine (IOM) committee, as well as 20 additional presentations to the committee's three working groups on medical-surgical priorities, health care, and behavioral, social, and cultural issues.

The subject matter covered during the 2-day visit included selection and retention of astronauts, biological effects of space radiation, telemedicine, family supports, postflight rehabilitation, evaluation and validation of countermeasures, risk management, efforts to develop artificial gravity, and many others. The committee's visit also included a luncheon discussion and sampling of food provided to astronauts during space travel and visits to space shuttle training facilities, the Flight Medicine Clinic, the Sonny Carter Neutral Buoyancy Facility, and mock-ups of portions of the International Space Station (ISS).

Agenda
NASA Johnson Space Center, Houston, Texas, February 22–24, 2000

Welcome to Johnson Space Center
David Williams

Opening Remarks and Introduction of IOM Committee
John R. Ball

Overview of Space and Life Sciences
David Williams

Space Shuttle Clinical Experience
Roger Billica

Phase 1 *Mir*, Clinical Experience
Thomas Marshburn

Analog Environment, Clinical Experience
Roger Billica

Longitudinal Study of Astronaut Health
Sam Pool

An Evidence-Based Approach for Space Medicine
David Williams

Ethical Issues in Space Flight
Baruch Brody, Ellen Baker, Jerry Homick, and Charles Sawin

Behavioral Health and Performance
Christopher Flynn and Albert Holland

Flight Medicine Clinic
David Dawson and Denise Baisden

International Space Station (ISS) Mock-ups
Terrance Taddeo and Craig Fischer

Pathophysiology of Disease in Microgravity
Thomas Marshburn

Pharmacotherapeutics in Microgravity and Clinical Concerns
Lakshmi Putcha and William Norfleet

Biological Effects of Space Radiation
Jeffery Jones

Resuscitation and Stabilization in Microgravity
Airway Management — William Norfleet
Cardiac Resuscitation — Tom Marshburn
Surgical Technique — David Williams

Astronaut Selection and Retention—Medical Approach
Smith Johnston and Roger Billica

Problem-Oriented Approach to Space Medicine

Clinical Care Requirements	David Dawson
Shuttle Orbiter Medical System Kits	Paul Stoner
Crew Health Care System	Terrance Taddeo
Initial Training	Philip Stepaniak
Skill Retention	Thomas Marshburn
Medical Informatics	Patrick McGinnis
Telemedicine	James Logan
Routine Health Assessment	Rainer Effenhauser
Extravehicular Activity	William Norfleet
Medical Sciences Laboratories	William Paloski

Postflight Rehabilitation
Beth Shepherd

Countermeasure Evaluation and Validation
William Paloski

Wednesday, February 23, 2000

Risk Management
Roger Billica

Countermeasure Evaluation: Extended-Duration Orbiter Medical Program
Charles Sawin

Critical Path for Countermeasure Development
John Charles

Clinical Care Capability Program
Craig Fischer

National Space Biomedical Research Institute and Countermeasure Development
Bobby Alford and Larry Young

Bioastronautics
David Williams

Food Lab, Food, and Nutrition
Helen Lane

Mission Architecture for Exploration-Class Missions
John Charles

Artificial Gravity
Sam Pool

Special Topics in Behavioral Health and Performance

Space Flight and Analogue Experience	Albert Holland
Psychological Adaption	Christopher Flynn
Human-to-System Interface	
Sleep and Circadian Assessment	

National and International Strategies for Space Medicine

Phase 1 Mir *Experience*	
Organization and Management	Sam Pool
Credentialing, Education, Training	Terrance Taddeo
Medical Certification of Crew	Roger Billica

International Space Station (ISS) Activities

Multinational Medical Operations Panel	Roger Billica
Multinational Space Medicine Board	Sam Pool
Multinational Medical Policy Board	Richard Williams (NASA Headquarters)
Human Research Review Board	Charles Sawin Lawrence Dietlein

Space Medicine and Space Exploration
David Dawson and Roger Billica

Integration of Ground-Based Medical Care

Introduction to Policy	David Williams
Flight Medicine Clinic	Denise Baisden and David Dawson
Training	Philip Stepaniak
Aerospace Medicine Board	Craig Fischer and Consultants
Medical Informatics	Patrick McGinnis
Rehabilitation Facility	Beth Shepherd

Star City (Russia) and Medical Care
Roger Billica

Emergency Medical Service Support for Shuttle Launch and Landing
Craig Fischer and David Dawson

Benefits to Terrestrial Health Care
David Williams

ADDITIONAL COMMITTEE VISITS TO JOHNSON SPACE CENTER

Members of the committee met with Capt. James Weatherbee and Dr. Steven Hawley of the Flight Control Operations Directorate and with Dr. Albert Holland of the Life Sciences Directorate on May 11, 2000, to obtain information regarding the process for selection of astronauts for space missions and with Drs. David Williams and Bobby Alford on June 22, 2000, to discuss NASA's organizational structure and the role of the National Space Biomedical Research Institute in reference to space medicine.

DENTAL WORKSHOP

The committee held an invitational workshop, Space Dentistry: Maintaining Astronauts' Oral Health on Long Missions, in Washington, D.C., on April 17, 2000. In planning the workshop, the committee solicited program ideas and the names of potential speakers from the dental practice and dental research communities. At the committee's request, the American Dental Association (ADA) helped secure a moderator for the workshop. The committee invited 20 dental researchers, dental practitioners, and materials scientists to attend the workshop and contribute to the information-gathering session. The workshop program and list of participants appear below.

The purpose of the workshop was to examine a variety of oral health issues in the context of space travel beyond Earth orbit. The committee was interested in learning of recent developments and whether dental disease or dental trauma, or both, are likely to emerge as problems on long missions. If they are, the committee wanted to know the available treatment options. The committee also wanted to hear updates on caries prevention; the linkage between diet and dental disease; oral hygiene in space; and progress on developing instruments, materials, and equipment suitable for use in a cramped and weightless environment.

Space Dentistry: Maintaining Astronauts' Oral Health on Long Missions Monday, April 17, 2000, Institute of Medicine Agenda

Welcome and Opening Remarks

John R. Ball, M.D., J.D., Chair, IOM Committee on Creating a Vision for Space Medicine During Travel Beyond Earth Orbit

Introduction

Colonel Shannon E. Mills, D.D.S. (Program Moderator), U.S. Air Force Dental Corps, USAF Inspection Agency, Kirtland AFB, New Mexico

Effects of Hypogravity on Calcified Tissue

Adele L. Boskey, Ph.D., Director of Research, Hospital for Special Surgery, New York, New York

Dental Caries: Initiation, Progression, and Prevention

Irwin Mandel, D.D.S., Professor Emeritus, Columbia University School of Dental and Oral Surgery, New York, New York

Managing Dental Trauma in Microgravity

Commander James C. Ragain, Jr., D.D.S., Ph.D., Dental Corps, U.S. Navy, Naval Dental Research Institute, Great Lakes, Illinois

Atraumatic Restorative Treatment (ART): A Potentially Useful Approach to Managing Dental Lesions in Space

Saskia Estupiñán-Day, D.D.S., Regional Oral Health Adviser, Pan American Health Organization, Washington, D.C.

Open Discussion

Committee members, program participants, and invited guests

Acknowledgments

American Academy of Periodontology
American Dental Association
International and American Associations for Dental Research
National Institute of Dental and Craniofacial Research
Pan American Health Organization

Workshop Guests

Dr. A. David Brandling-Bennett
Pan American Health Organization

Col. Gerry Caron
Office of the Surgeon General, U.S. Air Force

Dr. Robert J. "Skip" Collins
International and American Associations for Dental Research

Dr. Ray Dionne
National Institute of Dental and Craniofacial Research

Dr. Fred Eichmiller
American Dental Association Health Foundation, Paffenbarger Research Center

Dr. Isabel Garcia
National Institute of Dental and Craniofacial Research

Ms. Elise Handelman
Occupational Safety and Health Administration

Dr. Joseph G. Handelman
Private Practice, Annapolis, Maryland

Dr. Herschel Horowitz
National Institute of Dental and Craniofacial Research

Mr. Jonathan McLeod
American Dental Association

Dr. David Metzler
American Academy of Periodontology

Mr. Trevor Milner
Pan American Health Organization

Dr. Russell B. Rayman
Aerospace Medical Association

Dr. Gordon Rovelstad
American College of Dentistry

Dr. Gary Schumacher
ADA Health Foundation, Paffenbarger Research Center

MEDICAL CARE DURING SPACEFLIGHT: VIEWS OF PHYSICIAN-ASTRONAUTS WORKSHOP

Panel Discussion
Erik Johnson Conference Center, The National Academies,
Woods Hole, Massahcusetts,
July 27, 2000

In 30 years in space, only a handful of health care professionals have found themselves in a position to observe directly the symptoms of ill health in microgravity and to treat patients (including themselves) in a spacecraft. They are veteran astronauts who also happen to be physicians. One of them, Bernard A. Harris, M.D., is a member of the IOM committee that prepared this report. A retired astronaut and veteran of two shuttle missions, Dr. Harris helped organize a panel of one current and four former physician-astronauts. The purpose was to give committee members an opportunity to discuss firsthand issues of astronaut health and care during a space mission from the perspectives of astronauts who are also physicians.

Each of the physician-astronauts discussed with the committee their views on health care priorities for extended missions as well as a "top 10" list of astronaut health issues.

Bernard A. Harris, Jr., M.D., Moderator
SPACEHAB, Inc.
Houston, Texas

F. Andrew (Drew) Gaffney, M.D.
Vanderbilt University Medical Center

Chiaki Mukai, M.D.
Japanese Space Agency

Rhea Seddon, M.D.
Vanderbilt University Medical Center

Norman Thagard, M.D.
Florida State University

HEALTH CARE PLANNING FOR REMOTE HOSTILE ENVIRONMENTS WORKSHOP

J. Erik Jonsson Woods Hole Center, The National Academies, Woods Hole, Massachusetts, July 28, 2000

The workshop covered planning for missions; evaluation of treatment priorities; selection of personnel; crew training; and selection of supplies and equipment for long-duration submerged submarine, Antarctic, and North Sea and North Slope environments

Invited discussants were as follows:

Professor David H. Elliott, OBE, D. Phil.
Advisor to the Royal Navy and the International Marine Contractors Association, Surrey, United Kingdom

Commander Wayne G. Horn, M.D., USN
Director, Research and Development
Naval Submarine Medical Research Laboratory

Desmond J. Lugg, M.D.
Head, Polar Medicine
Australian Antarctic Division

Harry Mahar, Ph.D.
Antarctic Sciences Section, National Science Foundation

Lawrence A. Palinkas, Ph.D.
Department of Family and Preventive Medicine
University of California at San Diego

Background Paper for the Health Care Planning for Remote Hostile Environments Workshop (prepared by IOM Staff to the Committee)

Space may be the ultimate hostile environment, and astronauts traverse it in isolation and confinement. Protection of their well-being and preservation of their efficiency are of paramount concern during space missions. Delivery of medical care is rendered difficult, however, by weightlessness, limited equipment (because of payload restrictions), and the lack of information about the short- and long-term effects of exposure of humans to zero gravity. Experiments and data collection in studies with astronauts has been hampered by the small numbers of participants, the short durations of the studies, and the lack of available pre- and postflight data. There are numerous indications that some of the variability in findings may be due to genetic differences, but so far, no genetic studies have been undertaken.

At present, there is no way of simulating zero gravity for prolonged periods on Earth, but there is the opportunity to study small and large populations of volunteers in isolated, confined, and sometimes hostile environments. The planning procedures, personnel screening and training procedures, and equipment selected and used in various situations in analog environments can be used to prepare for long-duration space missions beyond Earth orbit, as well as to study common effects on physiology and behavior. Such studies may increase understanding of the changes produced by factors independent of the effects of zero gravity, such as loss of the normal diurnal cycle, sleep disorders, immune system changes, and changes in calcium metabolism. Health care problems in analog environments that appear to be most relevant for the Committee on Creating a Vision for Space Medicine During Travel Beyond Earth Orbit are the Antarctic experience, extended underwater submarine missions, and remote oil exploration activities.

Australian National Antarctic Research Expeditions (ANARE) has been collecting and analyzing data for more than 40 years under the guidance of

Desmond Lugg, and numerous publications on those experiences are available. Since 1993, ANARE has been engaged in cooperative international research and has recently forged alliances with NASA researchers. ANARE's organization, length of experience, ability to counter the problem of studies with small numbers of participants, and control of the isolation, confinement, and stress inherent in its experimental community make ANARE a valuable resource. ANARE has identified a number of health care issues that may be of interest to the committee.

It has long been assumed that stress causes alterations in the immune system. Studies with volunteers in confinement in Antarctica show an impaired immune response, an almost 50 percent reduction in T-cell proliferation after a challenge with phytohemagglutinin, and increased shedding of herpesviruses and Epstein-Barr virus. Researchers at Baylor University also found increased shedding of viruses in astronauts, but they noted more shedding before spaceflight, leading them to conclude that there was greater stress before flight than during or after the flight. ANARE noted a difference in immune impairment between individuals on two Antarctic stations, with the individuals at the more isolated station (with less communication and evacuation potential) showing a higher incidence of defects. Impairment of neutrophil function has also been detected in deep-sea divers. To date, none of the members of these isolated groups have suffered any recognizable immune-related infectious disease problems.

Alterations in vitamin D metabolism related to the absence of ultraviolet type B (UV-B) radiation have been noted in submariners as well as in groups who have spent the winter in Antarctica, but there is no evidence of decreased bone mass over experimental periods up to 1 year. There has been a greater than expected incidence of renal calculi among members of the ANARE group, but not among submariners, who were observed over a much shorter period. Both groups are sufficiently large to support controlled clinical studies.

Numerous observations of changes in hormone levels have been made, with documented changes in the pituitary-thyroid axis, named the polar T3 syndrome. Changes in parathyroid hormone (PTH) levels have been noted during space travel and may be related to bone mineral density loss. Furthermore, Antarctica, like space beyond Earth orbit, lacks an ozone layer to filter UV radiation; this may offer another avenue for research.

One would expect behavioral and psychiatric problems to be significant in the isolation and confinement of a 30-month mission. Hence, the ability to study the volunteers before, during, and after a winter in Antarctica would be valuable. Some studies have been done and published by ANARE (by J. A. Wood, D. J. Lugg and colleagues) and by Lawrence Palinkas at the Naval Health Research Center, among others. Although sleep disorders, anxiety, depression, irritability, fatigue, and several other symptoms are common, they very rarely impair function with any significance and are often transient. High levels of individual motivation and careful screening may be responsible for the minimization of symptoms. There is considerable variation in individual behavioral responses, with minimal impairment of group function. The implications of these and future studies are of great importance to long-duration space missions.

Submariners have been confined below the ocean's surface for periods of up to 3 months or more. Environmental changes include the absence of sunlight (and UV-B radiation), increased carbon dioxide levels (two to six times the normal level) with accompanying mild respiratory acidosis, and dietary restrictions. Dlugos and colleagues (Dlugos et al., 1995) found a decrease in the levels of urinary calcium excretion and increased PTH levels, along with other changes that seemed to protect against the formation of renal calculi. However, they caution that individuals who were known to form renal calculi were rejected from the missions, which may limit the comparability of submarine missions to spaceflight. Further data must be collected to assess the effect of elevated PTH levels on bone resorption, since no clinical effect was noted. There are many other areas in which controlled clinical studies will provide valuable information for space medicine.

Finally, experience in another analog environment, diving in the North Sea, is unique in that it is directed and financed by commercial organizations. Occupational health among divers is of paramount importance, and David Elliott and coworkers are experts on compression and decompression injuries and physiology, as well as immersion problems. The rigorous screening of divers is appropriate for their tasks, and the planning and delivery of medical care have been given high priorities.

The patterns of illness and injury among individuals in these diverse analog environments show considerable similarities, with some variations caused by methods of reporting, the durations of the missions, and the completeness of the medical studies. Roger Billica, chief of NASA's Medical Operations Branch, has noted that the Antarctic experience bears the closest resemblance to spaceflight, and emphasis will be placed on using data and information from that experience to prepare for flight beyond Earth orbit.

The examples of studies in analog environments noted above provide only a portion of the information already available, and there is the potential to acquire much more. Further cooperative studies can be done with greater safety, control, and follow-up than those available for spaceflight studies alone.

National Aeronautics and
Space Administration
Office of the Administrator
Washington, DC 20546-0001

NASA

JAN 5 1999

Kenneth I. Shine, M.D.
President
Institute of Medicine
2101 Constitution Avenue, NW
Washington, DC 20418

Dear Dr. Shine:

It was a great pleasure to meet with you at the Baylor College of Medicine in November. I am very pleased with the progress of the STS-88 mission, marking the beginning of construction of the International Space Station (ISS). We are underway to establish a permanent human presence in low-Earth orbit. There remains, however, a multitude of questions to be answered and problems to be solved for our explorers and international crews to safely live and work in the extreme environment of space beyond low-earth orbit, using the ISS as a research platform to prepare for the exploration of the solar system.

Space is arguably the most extreme environment that humans have ever entered. Prolonged presence in microgravity results in a number of adaptations that may not be completely reversible. Space crews are isolated, and spacecrafts provide limited room in which to live and work. Future exploration of the solar system also requires crews to land on other planets (such as Mars) and perhaps establish colonies on the Moon. Currently, radiation protection in deep space missions is an unsolved issue, and the threat of dysbarism is ever present, especially during extravehicular activities.

There are notable efforts underway to develop physiological countermeasures to maintain crew health and allow a rapid and uneventful readaptation upon return to Earth. The extramural Principal Investigators, funded under the NASA National Research Announcement grants system and the National Space Biomedical Research Institute, are two extremely talented groups that are addressing these issues. Efforts to develop a more capable medical-care delivery system in space, however, have largely been internal to NASA. We have traditionally depended on a preventive approach to maintain health, reflected in strict astronaut-selection standards and close monitoring of astronaut-health status. This has proven adequate for short-duration flight, and our crew has been essentially free of major illness during flight. This may not be the case in the future.

Your assistance is requested in evaluating our current medical-care system and recommending the type of infrastructure we will need to develop to support long-duration missions, including interplanetary travel in which timely evacuation of crew members will not be an option. Medical-care-provider training, specialty mix, nonmedical crewmember skills, use of advanced technology, surgical/intensive care capability in space, rehabilitation approaches to cope with exposures to gravitational fields following exposures to microgravity, psychological/human-factors challenges and use of robotics for health monitoring, education, and possible surgery are examples of the types of issues we would like you to address. We would also like you to consider the use of analog environments, such as remote Antarctic stations, for training and research. Ethical considerations in the face of limited medical-care capability are also important issues that need examination.

You and your colleagues will surely find this an interesting and stimulating project in which your efforts will prove critical to the overall success of long-duration human space flight. I have designated Dr. Arnauld E. Nicogossian, Associate Administrator for Life and Microgravity Sciences and Applications, as the NASA contact point in this matter. We look forward to engaging the Institute of Medicine in this important task.

Sincerely,

Daniel S. Goldin
Administrator

B
Committee and Staff Biographies

COMMITTEE BIOGRAPHIES

JOHN R. BALL, M.D., J.D. (*Chair*), is a member of the Institute of Medicine (IOM) and chaired the IOM Committee on Evaluating Clinical Applications of Telemedicine (*Telemedicine: A Guide to Assessing Telecommunications in Health Care*, National Academy Press, 1996). Dr. Ball is a graduate of Emory University, received a J.D. and an M.D. from Duke University in 1972, and was a Robert Wood Johnson Clinical Scholar at George Washington University from 1977 to 1979. After a residency in internal medicine at Duke Medical Center he held several health policy positions within the U.S. Department of Health and Human Services and was a senior policy analyst in the Office of Science and Technology Policy in the Executive Office of the President from 1978 to 1981. From 1986 to 1994 Dr. Ball was executive vice president of the American College of Physicians, and from 1995 to 1999 he served as president and chief executive officer of the Pennsylvania Hospital. Most recently, Dr. Ball was senior vice president and practice director of The Lewin Group. He is also a member of the American Clinical and Climatological Association and the Society of Medical Administrators.

JOSEPH V. BRADY, Ph.D., received a Ph.D. in behavioral biology from the University of Chicago in 1951, after which he served as chief of the Department of Experimental Psychology at the Walter Reed Institute of Research. From 1963 to 1970 he was the deputy director of the Division of Neuropsychiatry at that institution while also directing the Space Research and Psychopharmacology Laboratories at the University of Maryland. In 1967 he was appointed professor of behavioral biology and director of the Behavioral Biology Research Center at The Johns Hopkins University School of Medicine, a position he holds today. Dr. Brady has extensive experience in many areas of human behavior and has previously served on the National Research Council Committee on Space Biology and Medicine and on the Space and Earth Sciences Advisory Committee of the National Aeronautics and Space Administration.

BRUCE M. COULL, M.D., is professor of neurology and head of the Department of Neurology at the College of Medicine of the Arizona Health Sciences Center. From 1982 to 1995 he served as director of the Comprehensive Stroke Center of Oregon at the Oregon Health Sciences Center, and from 1991 to 1995 he was professor of neurology at the Oregon Health Sciences Center. Dr. Coull is on the editorial boards of four major journals dealing with stroke and related issues. He is a member of the Executive Committee of the American Heart Association's Stroke Council and has served as the principal investigator or a co-principal investigator on numerous studies of stroke, including the multicenter Stroke Prevention in Atrial Fibrillation (SPAF I, II, and III) study, the National Institute of Neurological Disorders and Stroke-funded Mechanism of Injury and Repair in Ischemic Stroke study, and the North American Symptomatic Carotid Endarterectomy Trial.

N. LYNN GERBER, M.D., graduated from Tufts University Medical School in 1971 and completed a residency in internal medicine at the New England Medical Center, followed by a fellowship in rheumatology at the National Institutes of Health. She then took a residency in physical medicine and rehabilitation at George Washington University, which she completed in 1977. Dr. Gerber is currently chief of the Rehabilitation Department of the Clinical Center of the National Institutes of Health and holds faculty appointments at Georgetown and George Washington Universities. She is board certified in internal medicine, rheumatology, and physical medicine and rehabilitation. She is widely published and has received numerous awards. Much of her work involves the study of mechanics of foot function, joint preservation, and energy conservation in individuals with rheumatoid arthritis, and she has helped develop programs for the use of ultrasound imaging of the musculoskeletal system.

BERNARD A. HARRIS, Jr., M.D., is a retired astronaut who flew two missions of more than 438 cumulative hours, with 4 hours of extravehicular activity, during his 6-year tenure (1990–1996) in the astronaut corps. He received an M.D. from Texas Tech University Health Sciences Center in 1982 and completed an internal medicine residency at the Mayo Clinic in 1985. He held a National Research Council endocrine fellowship for 2 years and was awarded an M.M.S. by the University of Texas Medical Branch, Galveston, and an M.B.A. in 1999 by the University of Houston. Dr. Harris is an associate professor of internal medicine at the University of Texas Medical Branch, Galveston, and served as director of its Center for Aerospace Medicine and Physiology. He also served as the chief scientist and vice president for science and health services at Spacehab, Inc., which designs and builds laboratory modules for the National Aeronautics and Space Administration (NASA). Dr. Harris is a member of the National Institutes of Health National Institute on Deafness and Other Communication Disorders Advisory Council, the NASA Life and Microgravity Sciences and Applications Advisory Committee, and the NASA Life Science Subcommittee.

CHRISTOPH R. KAUFMANN, M.D., M.P.H., is a 1982 graduate of the Uniformed Services University of the Health Sciences (USUHS) who completed a fellowship in trauma and critical care and received an M.P.H. at the University of Washington after completion of his surgical training. He is currently an associate professor of surgery at USUHS, where he directs the Trauma Program and serves as principal investigator on four major grants. He is heavily involved in computer simulation training and telesurgery and has planned and will run a major training laboratory for students and staff. Col. Kaufmann is active in local and national trauma organizations, site visits, and editorial review and regularly contributes to the literature on trauma and military medicine. He served on an Institute of Medicine exploratory committee on training for new technology.

JAY M. McDONALD, M.D., received an M.D. from Wayne State University in 1969. After an internship in internal medicine at the University of Oregon, he returned to Wayne State University to complete a residency in pathology in 1974. He then moved to Washington University for a postdoctoral research fellowship and remained on the faculty until 1990. He was the director of the Division of Laboratory Medicine from 1980 to 1990. He then moved to the University of Alabama, Birmingham, where he assumed the chair of the Department of Pathology. Dr. McDonald's research interests include diabetes, metabolism, osteoporosis, and calcium metabolism, and he has served on numerous study groups and organizations related to these topics. Dr. McDonald has been a prolific contributor to the scientific literature and serves on the editorial boards of three major journals.

RONALD D. MILLER, M.D., received an M.D. from Indiana University in 1964 and had a residency in anesthesiology at the University of California, San Francisco, where he was also awarded an M.S. in pharmacology in 1967. He remained at the University of California, rising to the position of professor and chair of the Department of Anesthesia and professor of molecular and cellular pharmacology in 1984. He has assumed leadership positions in many local and national organizations involved with anesthesia, pharmacology, and critical care and has chaired the National Institutes of Health Surgery, Anesthesia, and Trauma Study Section and the Food and Drug Administration Anesthesia and Life Support Drug Advisory Committee, among many others. He has authored 54 book chapters and books, including the most widely used anesthesia text, and approximately 200 original articles. Dr. Miller is an Institute of Medicine (IOM) member who has served on several IOM committees.

ELIZABETH G. NABEL, M.D., graduated from Cornell University Medical College in 1981 and then completed a residency in internal medicine at Brigham and Women's Hospital. She spent the next 4 years as a clinical and research fellow at that institution before moving to the faculty at the University of Michigan. She was appointed director of the Cardiovascular Research Center in 1992, professor of medicine in 1994, and professor of physiology a year later. In 1997 she was named chief of the Division of Cardiology at the University of Michigan, and in September 1999 was appointed director of clinical research at the National Heart, Lung, and Blood Institute of the National Institutes of Health. Dr. Nabel has been the principal investigator on numerous research grants and has been an active teacher at several academic levels. She is a member of the Institute of Medicine.

TOM S. NEUMAN, M.D., is professor of clinical medicine, director of the Hyperbaric Medicine Center, and attending physician in the Pulmonary Division at the University of California San Diego Medical Center. A graduate of Cornell University, he received an M.D. from the New York University School of Medicine in 1971, followed by performance of an internship and a residency in internal medicine at Bellevue Hospital. Dr. Neuman is board certified in internal medicine, pulmonary disease, occupational medicine, and emergency medicine. He is a fellow of the American College of Physicians and the American College of Preventive Medicine. Dr. Neuman has been a leader in the field of the physiology and medicine of diving throughout his career and is the editor-in-chief of *Undersea and Hyperbaric Medicine*.

DOUGLAS H. POWELL, Ed.D., is a clinical psychologist. He holds a diplomate in clinical psychology. He received his doctorate from Harvard University and has remained affiliated with that institution for 40 years. He has been a senior clinician at Harvard University Health Services and has been chief of the Psychology Service, director of Training and Research, and coordinated the Behavior Therapy program. Dr. Powell has held academic appointments on the faculties of psychology, extension, and, currently, medicine. He has had extensive contract with members of the university on many levels: assessment, counseling, advising, and teaching. While in the U.S. Air Force Dr. Powell evaluated astronaut candidates for the Gemini and Apollo programs. His four books focus mainly upon varieties of normal human development from the adolescent to the young-old years, and he is the senior author of MicroCog, a computerized neuropsychological screening test. He has had a career-long interest in the prediction of behavior, selection, and training. He is a senior partner at Powell and Wagner Associates, a Cambridge, Massachusetts-based consulting firm.

WALTER M. ROBINSON, M.D., M.P.H., received an M.D. from Emory University in 1988, after which he spent 2 years at Boston City Hospital and 1 year at The Johns Hopkins University as a pediatric resident. After working in a neighborhood health center for 1 year, he returned to a fellowship in pediatric pulmonary medicine at Harvard/Children's Hospital and also served a fellowship in medical ethics. He received an M.P.H. from the Harvard School of Public Health in 1994 and completed a fellowship in the Program in Ethics and the Professions in 1998. Dr. Robinson is board certified in pediatrics and pediatric pulmonary medicine. He has continued to practice while teaching ethics at the medical school and postgraduate levels and serving on committees appropriate to his expertise. He is the associate director of the Division of Medical Ethics and directs the Harvard Medical School Ethics Fellowship Program, as well as the Program in the Practice of Scientific Investigation, and serves on the editorial board of *Ethics and Behavior*.

CAROL SCOTT-CONNER, M.D., Ph.D., received an M.D. from the New York University School of Medicine in 1976 and stayed there for her surgical residency, which she completed in 1981. After leaving New York University, she joined the faculty at Marshall University and then moved to the University of Mississippi. During her tenure there she earned a Ph.D. in anatomy from the University of Kentucky and an M.B.A. Since 1995, she has been professor and head of surgery at the University of Iowa. Dr. Scott-Conner has been active on 22 editorial boards and has authored more than 200 original papers, abstracts, reviews, and book chapters. She holds memberships in many elected surgical societies and has frequently served in leadership positions.

JUDITH E. TINTINALLI, M.D., received an M.D. from Wayne State University in 1969 and immediately entered a residency in internal medicine at Detroit General Hospital, which she completed at the University of Michigan. While working as the clinical director of the Department of Emergency Medicine at William Beaumont Hospital, she received an M.S. in biostatistics and research design from the School of Public Health of the University of Michigan. Dr. Tintinalli is chair and residency program director of the Department of Emergency Medicine at the University of North Carolina. She is board certified in internal medicine and emergency medicine and has served as president of the American Board of Emergency Medicine as well as chair of its Research Committee and member of the Test Development Committee. She has represented the group at the American Board of Medical Specialties. She has taken a leadership position in many organizations related to her specialty and is involved in community and governmental service. Dr. Tintinalli serves on several editorial boards, is widely published, and is a member of the Institute of Medicine.

Liaison to the Board on Neurosciences and Behavioral Health

STEVEN M. MIRIN, M.D., is medical director of the American Psychiatric Association in Washington, D.C., a medical specialty organization with approximately 40,000 physician members in 76 district branches nationwide. He was president and psychiatrist in chief of McLean Hospital in Belmont, Massachusetts. Dr. Mirin is professor of psychiatry at Harvard Medical School. He has also served as president of the Massachusetts Psychiatric Society, chair of the Governing Council of the Section for Psychiatric and Substance Abuse Services of the American Hospital Association, and co-editor-in-chief of the *Harvard Review of Psychiatry*. His research activities have focused on the biological and psychosocial aspects of substance use disorders and the outcomes of care for psychiatric patients generally, for which he received the Presidential Award for Research from the National Association of Psychiatric Health Systems in 1991.

Liaison to the Space Studies Board

MARY JANE OSBORN, Ph.D., has a primary research interest in the biogenesis of bacterial membranes. Dr. Osborn was a research associate at the New York University College of Medicine from 1959 to 1961 and an assistant professor from 1961 to 1962. She progressed from assistant to associate professor at the Albert Einstein College of Medicine in New York (1963–1968). In 1968, she joined the University of Connecticut Health Center as a professor in the Department of Microbiology. Dr. Osborn has been head of the department since 1980. Dr. Osborn has served on the President's Committee on the National Medal of Sciences from 1981 to 1982, as a member of the National Science Board from 1980 to 1986, and as a member of the Advisory Council of the National Institutes of Health's Division of Research Grants (1989–1994), for which she was chair from 1992 to 1994. She has also served on the Council of the American Society of Biological Chemists from 1974 to 1977 and, again, as president from 1981 to 1982; as chair of the American Chemical Society's Division of Biological Chemistry (1975–1976); and as president of the Federation of American Societies for Experimental Biology (1982–1983). Dr. Osborn is a member of the National Academy of Sciences (1978), the American Academy of Arts and Sciences, and the American Academy of Microbiology (1992). Dr. Osborn is also the past chair of the Board's Committee on Space Biology and Medicine.

Liaison to the Board on Health Sciences Policy

GLORIA E. SARTO, M.D., Ph.D., is professor and past chair of the Department of Obstetrics and Gynecology at the University of New Mexico

School of Medicine in Albuquerque. Her research interests include studies of genetic disorders and reproductive dysfunction. Dr. Sarto is president of the Society for the Advancement of Women's Health Research and is on the Professional Advisory Board of the Epilepsy Foundation of America. She is a member of the Board of Governors and Board of Directors of the National Center for Genome Resources and chairs the Advisory Council for obstetrics and gynecology of the American College of Surgeons. Dr. Sarto was a member of the National Advisory Council on Child Health and Human Development, National Institutes of Health (NIH); the Clinical Research Panel of the National Task Force on the NIH Strategic Plan; and the Committee on Research Capabilities of Academic Departments of Obstetrics and Gynecology, Institute of Medicine, National Academy of Sciences. Additionally, she has been vice president of the American Board of Obstetrics and Gynecology and director of its Division of Maternal-Fetal Medicine; a grant reviewer for the NIH Reproductive Biology Study Section and Human Embryology and Development Study Section; a panel member for the NIH Consensus Development Conference on Cesarean Childbirth; and a panel member for Treatment Effectiveness of Hysterectomy and Other Therapies for Common Noncancerous Uterine Conditions, a conference of the Agency for Health Care Policy and Research, Center for Medical Effectiveness Research. Dr. Sarto has published extensively on a wide array of women's health topics, including reproductive medicine and sexually transmitted diseases. She currently is on the editorial boards of *Perinatal Press*, *Journal of Reproductive Medicine*, and *Women's Health Letter*.

IOM PROJECT STAFF

CHARLES H. EVANS, Jr., M.D., Ph.D., is Senior Adviser for Biomedical and Clinical Research at the Institute of Medicine of the National Academies. A pediatrician and immunologist, he graduated with a B.S. in biology from Union College and an M.D. and a Ph.D. from the University of Virginia and trained in pediatrics at the University of Virginia Medical Center. From 1975 to 1998 he served as chief of the Tumor Biology Section at the National Cancer Institute and holds the rank of Captain (Ret.) in the U.S. Public Health Service with 27 years of service as a medical scientist at the National Institutes of Health in Bethesda, Maryland. Dr. Evans's research interests include carcinogenesis (the etiology of cancer), the normal immune system defenses to the development of cancer, and aerospace medicine. He discovered the ability of cytokines to directly prevent carcinogenesis and was the first to isolate a direct-acting anticarcinogenic cytokine, for which he was awarded four U.S. patents. Dr. Evans is the author of more than 125 scientific articles and is the recipient of numerous scientific awards including the Outstanding Service Medal from the U.S. Public Health Service and the Wellcome Medal and Prize. He is a fellow of the American Association for the Advancement of Science and the American Institute of Chemists and is a credentialed fellow in health systems administration of the American Academy of Medical Administrators. An active adviser to community medicine and higher education, he has served on the Board of Trustees of Suburban Hospital Health System and on the College of Arts and Sciences Board of Trustees at the University of Virginia. Dr. Evans is the study director for the Committee on Creating a Vision for Space Medicine During Travel Beyond Earth Orbit.

MELVIN H. WORTH, Jr., M.D., is a scholar in residence at the Institute of Medicine. Dr. Worth completed his surgery residency at New York University-Bellevue in 1961 and remained on that faculty for 18 years. He founded the Bellevue Trauma Service in 1966 and continued as director until 1979, when he left to become director of surgery at Staten Island University Hospital. He served for 15 years with the New York State Office of Professional Medical Conduct and 8 years as a member of the New York State Hospital Review and Planning Council (for which he was chair in 1993). He is a fellow of the American College of Surgeons, the American College of Gastroenterology, and the International Society for Surgery and holds memberships in the American Association for the Surgery of Trauma, the Society for Critical Care Medicine, the Association for Academic Surgery, the New York Surgical Society (for which he was president in 1979), and other academic and professional organizations. Dr. Worth retains his appointment at New York University and is clinical professor of surgery at the State University of New York Downstate (Brooklyn) and the Uniformed Services University of the Health Sciences. Dr. Worth most recently served as an Institute of Medicine study staff member to the Committee on Fluid Resuscitation for Combat Casualties and is senior adviser to the Committee on Creating a Vision for Space Medicine During Travel Beyond Earth Orbit.

JUDITH RENSBERGER, M.P.H., is a biomedical research and health policy analyst. She has 11 years of experience in government relations and 8 years of experience in health and science communications. From 1993 to 1999 Ms. Rensberger was manager, legislative and regulatory policy, for the

American Dental Association and from 1988 to 1993 served as senior health policy analyst for the National Committee to Preserve Social Security and Medicare. Before that, from 1985 to 1987 she was communications director at the Foundation for Biomedical Research. She has advanced degrees in public health and science communications, plus additional work toward a doctorate in epidemiology. Ms. Rensberger is the senior program officer for the Committee on Creating a Vision for Space Medicine During Travel Beyond Earth Orbit.

VERONICA A. SCHREIBER, M.A., M.P.H., received an M.P.H. with a concentration in epidemiology from the George Washington University in 1999. For her master's research project, she designed and conducted research on patient referral and consultation practices by physicians of the Ambulatory Care Department of the George Washington University Medical Center. Before her health research work, she held teaching positions in political science and international relations for 11 years at the University of the Philippines (1974 to 1980), University of Pittsburgh (1980 to 1983), Frostburg State University (1983 to 1984), and the University of Maryland European Division (1988 to 1989). She was a Ph.D. candidate at the University of Pittsburgh when she accepted a teaching position at Frostburg State University in 1983. She was research consultant at the Bonn office of the European Institute for Environmental Policy (1987 to 1988) and training specialist at the German Foundation for International Development (1988 to 1991). She is a member of the Pi Sigma Alpha Honor Society and the Phi Theta Kappa International Honor Society. Ms. Schreiber is the research assistant for the Committee on Creating a Vision for Space Medicine During Travel Beyond Earth Orbit.

SETH M. KELLY is a 1999 graduate of Dartmouth College, where he earned an A.B. degree in philosophy with a minor in public policy (biotechnology and health policy). During the spring of 1998 Mr. Kelly designed, administered, and analyzed the results of a survey of 16 hospices and home care facilities as part of the New Hampshire End of Life Project at Dartmouth Medical School, and from June to September of that year he was an environmental health intern with the Massachusetts Department of Public Health. He is a Wilderness Emergency Medical technician; served as medical officer, finance officer, and membership director of the Upper Valley Wilderness Response Team in Hanover, New Hampshire; and is a member of the National Association of Emergency Medical Technicians and of the American Society of Bioethics and Humanities. Mr. Kelly is a project assistant for the Committee on Creating a Vision for Space Medicine During Travel Beyond Earth Orbit.

TANYA M. LEE is a project assistant for the IOM Committee on a Strategies for Small Number Participants Clinical Research Trials and the Committee on Creating a Vision for Space Medicine during Travel Beyond Earth Orbit in the Board of Health Science Policy. She has been with the National Academies since April 2000. Tanya has attended the University of Maryland Eastern Shore and Prince Georges Community College, pursuing a degree in the field of sociology.

Index

A

Abscesses, 82, 84, 127
Accidents and injury, *see* Trauma
Aerospace Medical Association, 34
Age factors, 11, 58, 64, 77, 78
  breast cancer, 89
  medical events in space, 81
Agency for Healthcare Research and Quality, 205
Airway management, 111, 118-119
Alimentary system, *see* Gastrointestinal system
Ambient lighting, *see* Lighting in spacecraft
American College of Surgeons, 192
American Medical Association, 34
Ames Research Center, 193
Analog environments, 12, 13, 28, 30, 73, 80, 195, 249, 256-260, 262
  see also Antarctic stations; International Space Station
  behavioral issues, 12, 13, 139, 140-141, 142, 145, 149, 152, 159-160, 162, 164, 165-167, 170
    group interactions, 145-146, 152, 156, 159-160, 162, 164, 165-167
    mental illness, 80-81, 82, 85, 86
  bone mineral density loss, 40
  confidentiality issues, 177-178
  emergency and continuing care, 117-118, 126-127, 131, 132-133, 134
  gastrointestinal medical events, 85, 86
  infections, 82, 84, 85, 105, 126
  longitudinal studies, 82, 83-86
  microgravity, 76
  neurological effects, 53
  pharmacokinetics and pharmacodynamics, 91
  respiratory disorders, 82, 84, 85, 86
  submarines, 76, 80-83, 84, 85, 106, 140-141, 152, 162, 205, 256-257, 259
  surgery, 126-127
  urinary system, 82
Anemia, 39, 105, 106, 113
Anesthesia, *see* Pain management and anesthesia
Animal models
  behavioral, 150

bone mineral loss, 45-46
injury, responses to, 121
muscle loss, 48
nervous system, 53
rehabilitation measures, 127
surgical wounds, 121, 123
Antarctic stations, 12, 30, 76, 80, 83-86, 100, 195, 256-259
behavioral issues, 139, 140-141, 142, 159-160, 162
emergency and continuing care, 117-118, 126, 131, 132-133
infections, 105, 126
surgical wounds, 126, 127
Appendicitis, 29, 82, 85, 88, 102, 104, 127
Astronaut Medical Evaluation Requirements Document, 97
Ataxia, 39, 48
Atraumatic restorative treatment, 100
Attitudes and beliefs, *see* Public opinion/education
Auditory perception, 77
noise effects and controls, 8, 28, 93, 95
Australian National Antarctic Research Expeditions, *see* Antarctic stations
Authority, NASA organization, 20, 21, 192, 193, 194-195, 198-199
see also Leadership

B

Baroreceptors, 49, 50
Behavioral health, 28, 76, 137-172, 190, 249, 251, 262
see also Cultural issues; Group interactions; Mental health and illness; Mood effects; Screening, selection and training
analog environments, 12, 13, 139, 140-141, 142, 145, 149, 152, 159-160, 162, 164, 165-167, 170
Antarctic stations, 139, 140-141, 142, 159-160, 162
group interactions, 145-146, 152, 156, 159-160, 162, 164, 165-167
mental illness, 80-81
animal models, 150
clinical trials, 66
committee methodology, *xii*, 3
committee recommendations, 1, 11-13, 14, 15, 16, 21
computer applications, behavioral adaptation, 151, 154-155, 163-164, 165, 166
confined environments, 138, 140, 142, 168, 170
countermeasures, 107, 108-110, 142, 156-157, 164-165, 169
databases, 139, 142, 165-166
disruptive crew interactions, 14, 108, 154-155, 168, 170
end-of-life care, 9, 130, 134, 211
genetic factors, 162-163
ground support personnel, 108-109, 140, 164, 166, 168
health care personnel, 150
international authorities, cooperation on, 138, 140
international crews, 107, 137, 145, 170, 168
International Space Station (ISS), 150, 152-153, 159
medical events in space, 80-81
Mir space station, 140, 141, 153
NASA efforts, 138, 139, 141, 143, 146, 147, 148, 150-151, 154, 156-159, 160, 162, 164-165, 170
astronaut esprit as a problem, 174-175, 182-183
neurological factors, 162, 163
organizational factors, 20, 193
physiological monitoring of, 154, 155-156, 169
screening, 12, 77, 86, 107, 108
group interactions, 146-147, 148, 149, 158-165, 169, 170
underreporting of problems, 174-175, 183, 187
sexual behavior, 138, 142, 149; *see also* Contraceptives and contraception
sleep and circadian rhythm, 5, 39, 53-55, 64, 72, 78, 139, 142, 143-144

space crew/ground personnel interactions, 108-109, 140, 152-158, 166, 168
space shuttle program, 140, 155
Behavioral self-management, 145
see also Self-assessment; Self-monitoring
Bioastronautics Institute, 192, 196, 251
Bioavailabilities, 104, 113
Bioregenerative systems, 89, 143, 145
Blood, *see* Cardiovascular system; Hematology
Body mass, 90, 127, 128
Bone, *see* Musculoskeletal system
Bone mineral density, 3, 5, 9, 26-27, 39, 40, 42-47, 62, 64, 66, 77, 123, 127, 128, 129, 153
analog environments, 40
animal models, 45-46
biomarkers, 5, 43, 45, 46
bone resorption, 43, 259
calcium absorption, 43, 259
calcium excretion, 43, 60, 259
clinical research, 5, 46-47, 72
countermeasures, 27, 44, 46, 47, 60-61, 72, 127-128
gender factors, 43, 58, 59
genetic factors, 5, 43, 45, 47, 72
medications, 5, 44, 46, 47, 72
metabolic processes, 43, 44-46, 47, 60-61, 87, 259
Mir space station, 43, 44, 128
nutrition, 44
osteoporosis, 45, 47, 64, 212
real-time measurements, 46, 72
Bone resorption, 43, 259
Bowel sounds, 52, 87
Breast cancer, 86-89

C

Calcium absorption, 43, 259
Calcium excretion, 43, 60, 259
Carcinogenesis and cancer, 78, 86-89
breast, 86-89
gastrointestinal system, 103
prostate, 112
radiation effects, 65, 88-89, 212
Cardiovascular system, 8, 49-50, 77, 78
anesthesia and pain management, 118
central venous pressure, 49-51
clinical research, 5, 48-49, 50, 62, 64, 66, 72
end-of-life care, resuscitation considerations, 9, 130, 249
health care practice, 97-98
historical perspectives, 97
hydrostatic pressure, 39
injury, responses to, 121-122
medical events in space, 81, 113, 212
medical events in submarines, 82, 85
microgravity, 8, 49-50, 97
neurovestibular adaptation, 67
orthostatic hypotension, 5, 39, 48-49, 50, 97, 102
physical examinations in space, 87
renal system, vascular resistance, 50
screening, 97, 98, 113
Catastrophic illness and death, 6, 29, 114, 130, 138, 218
end-of-life care, 9, 130, 134, 211
historical perspectives, 78, 79
suicidal and homicidal ideation, 108, 154
triage, 118, 134, 211-212
Cell-mediated immunity, 105, 258
Chemicals, hazardous, *see* Hazardous substances
Central nervous system
anaesthesia, 120
closed-head injuries, 111
orthostasis, 110
psychomotor functions, 55
radiation effects, 65
sleep and circadian rhythm, 53
Central venous pressure, 49-51
Circadian rhythm, *see* Sleep and circadian rhythm
Clinical Care Capability Development project, 209, 250
Clinical practice guidelines, 8, 9, 31, 134, 204-205, 208, 217, 218
end-of-life considerations, 9, 130, 134

multiple casualties, 118, 134
rehabilitation measures, 127-128
standards, 7, 97, 113, 210-212

Clinical research, 3, 4, 6-7, 9-10, 24, 38, 40, 41, 62-73
see also Clinical Research Opportunities; Ethical issues
behavioral health, 66
bone mineral density, 5, 9-10, 42-43, 45, 46-47, 62, 64, 66, 72
cardiovascular effects, 5, 48-49, 50, 62, 64, 66, 72
committee methodology, *xii*, 3, 30, 31-32
Critical Path Roadmap project, 4, 62-69
defined, 69
gender factors, 5, 60, 72
hazardous substances, 65
hematology, 9, 62, 64, 106, 113
historical perspectives, 6, 10
immune system, 9, 62, 64, 66, 106, 113
infections, 9, 62, 64, 66
International Space Station (ISS), 41, 73
medications, 5, 55, 204
microgravity, 5, 6, 9-10, 17, 42-43, 45, 46-47, 62-69
musculoskeletal system, 5, 9-10, 42-43, 45, 46-47, 62, 64, 66, 72
neurological, 5, 10, 66, 72, 111, 113
neurovestibular function, 53, 62, 65, 67
nutrition, 5, 62, 63, 65
organizational requirements, 70, 193, 197-198, 199, 207-208
pharmacokinetics and pharmacodynamics, 5, 72, 90
physiological monitoring, 5, 61, 63, 68
radiation exposure, 28, 65
renal stones, 66
reproductive system, 5, 57, 59, 72
space motion sickness, 5, 52, 72
trauma, 63, 65
various body systems and processes, 5, 64-67, 72
wounds, 9, 106

Clinical Research Opportunities, 5, 72
bone mineral density, 5, 46-47, 72
gender factors, general, 60
nervous system, 5, 72
orthostatic hypotension, 5, 48-49, 72
physiological monitoring, 5, 62, 72
reproductive system, 5, 57, 59, 72
sleep and circadian rhythm, 5, 54-55, 64, 72
space motion sickness (SMS), 5, 52, 72

Clinical Research Plan, 202

Closed-head injuries, 111

Communication and communications technology, 2, 6, 11, 13, 142, 143, 151
see also Isolation
computer analysis of verbal communication, 154-155
emergency and continuing care, 118, 132, 204-205
family and friends, 14, 151, 152, 168
group interactions, 147, 154, 166
monitoring, 15, 151, 169
robotics, 24, 123-124, 262
surgical procedures, 123-124
time delays, 25, 109

Comprehensive health care system (NASA), 7, 9, 21, 40, 70-71, 190-191, 195, 196, 197, 198, 199, 200, 201, 208-209, 216, 218-219

Computer-aided design/computer-aided manufacturing (CAD-CAM), 124

Computer applications, other
see also Databases
behavioral adaptation, 151, 154-155, 163-164, 165, 166
informatics, 9, 24, 61, 208, 212-214, 217
recreational software, 151
robotics, 24, 123-124, 262
self-assessment software, 109
verbal communication analysis, 154-155

Confidentiality and privacy, 13, 16, 173, 174, 175-187, 195, 196, 201
analog environments, 177-178

crew onboard space vessels, privacy, 138, 142, 144
Confined environments, 26, 29, 189
anesthesia and pain management, 118
behavioral problems, 138, 140, 142, 168, 170
neurological effects, 53
privacy onboard spacecraft, 138, 142, 144
Contraception and contraceptives, 60, 104
Cosmic particles, 27, 96, 113
Counter Measure Development and Validation Project, 195-196
Countermeasures, 40-42, 250
see also Medications
behavioral problems, 107, 108-110, 142, 156-157, 164-165, 169
bone mineral density, 27, 44, 46, 47, 60-61, 72, 127-128
closed-head and spinal cord injuries, 111
Critical Path Roadmap project, 4, 62-69, 200, 211, 250
decompression sickness, 92
exercise, 48, 58, 127-128, 129, 142
hazardous materials exposure, 8, 94, 113
hormonal contraceptive agents, 60, 104
hormone replacement therapy, 9, 58-59, 104, 113
International Space Station (ISS), 41-42, 68
microgravity effects, 40-42, 48, 56, 62-69, 72, 127-128
artificial gravity, 5, 72, 251
muscle loss, 48, 56, 72, 127-128
NASA programs, general, 27, 68, 193, 195-194
behavioral, 156-157
bone mineral loss, 40-42, 46
noise effects and controls, 8, 28, 93, 95
organizational requirements, 195-196
peripheral nervous system, 56
recreation and leisure activities, 14, 128-129, 142, 144, 151
rehabilitation measures, 7, 8, 11, 28, 67, 71, 73, 127-130, 190, 250
surgical wound healing, 123
underreporting of problems, 175, 183, 187
urinary system, 60-61, 112
Crew selection and composition, *see* Ethnicity; International crews; Screening, selection and training
Crew Status and Support Tracker, 14, 168
Critical Path Roadmap project, 4, 62-69, 200, 211, 250
Cultural issues
see also Ethnicity; International crews; Public opinion
astronaut esprit as a problem, 174-175, 182-183
committee methodology, *xii*, 3
committee recommendations, 11, 21
ethical issues, 174-175, 182
food preparation 88
group interactions, 149, 156, 164-165, 169
language factors, 107, 140, 164-165, 169
mental health, 107, 149, 156, 158

D

Databases, 9, 11, 73
behavioral issues, 139, 142, 165-166
bone mineral density, genetics, 47
emergency and continuing care, 132-133, 134
pharmacokinetics and pharmacodynamics, 8, 91, 113
Decisionmaking, general
see also Ethical issues; Organizational factors
emergency and continuing care, 117, 118, 130, 134
group, 147
individual astronauts, 4, 21, 31, 194
informatics, 9, 24, 61, 208, 212-214, 217

leadership,
crew, 10, 108, 117, 110
NASA organization, 20, 21, 192, 193, 194-195, 198-199
triage, 118, 134, 211-212
Death and dying, *see* Catastrophic illness and death
Decompression sickness, 59, 67, 92-93
Dehydration, 51, 52, 60, 90, 111-112, 212
Dental care, 8, 84, 85, 98-101, 113, 252-255
gingivitis, 105
Department of Defense, 194, 200, 203, 204, 205
Department of Health and Human Services, *see* Agency for Healthcare Research and Quality; *terms beginning "National Institute..."*
Department of Veterans Affairs, 204, 205
Diagnostic and Statistical Manual, 159
Diarrhea, 102
Diet, *see* Nutrition
Digestive system, *see* Gastrointestinal system
Dual agency, 175
Dying, *see* Catastrophic illness and death
Dysbarism, 51, 189, 261

E

Electrical stimulation, 128, 129
Emergency and continuing care, 25, 30, 117-136, 169
see also Pain management and anesthesia; Surgery; Trauma
analog environments, 117-118, 126-127, 131, 132-133, 134
databases, 132-133, 134
decompression sickness, 92
microgravity, 121-123, 125, 127, 128, 129
NASA efforts, 126, 129-130, 134-135
organizational factors, 190, 192, 210-212
triage, 118, 134, 211-212
End of life considerations, *see* Catastrophic illness and death
Endocrine function, 258
contraceptives, 60, 104
exogenous hormone therapy, 9
gametes, female, 57
growth hormone, neurological effects, 5
injury, responses to, 121, 123
medical events in space, 81, 101
neurohormonal, 49
reproductive, 9, 57, 58, 72
replacement therapy, 9, 58-59, 104, 113
testosterone, 57
Endotracheal intubation, 118, 119
Environmental Protection Agency, 194
Epidemiological data, 1, 9, 59, 197
see also Analog environments; Clinical research; Longitudinal studies
Ergonomics, 95-96, 113
Erythropoietin, 40, 105
Ethical issues, 77, 173-187, 249
see also Confidentiality and privacy
committee methodology, *xii*, 30
flight surgeon, 174, 175, 180-181, 215
international crews, 174, 186
NASA efforts, 174, 175, 181-187
risk assessment, 4, 30, 31, 114, 206-207
public education, 6, 31, 114, 124, 131, 205-207
Ethnicity, 11, 13, 88
see also Cultural issues; International crews
Exercise countermeasures, 48, 127-128, 129, 142
female reproductive physiology, 58
Extravehicular activity (EVA), 77, 116, 189, 190, 191
decompression sickness, 92
gender factors, suits, 60
medical events in space, 79
space motion sickness (SMS), 52
Extreme environments, 1-2, 4, 6, 11-12, 13, 14, 28, 170, 189, 256-257
see also Analog environments; Confined environments; Isolation; Microgravity; Radiation exposure

group interactions, 145, 170
Mars exploration, 23, 24-25, 30
medical events, 78-86
space medicine defined, 34
Eye-hand coordination, 55, 125

F

Facial and periorbital edema, 26, 87, 90
Fatigue, 53, 55, 66, 106, 107, 153, 155, 212, 259
Federal Aviation Administration, 203-204
Females, *see* Gender factors
Food, *see* Nutrition
Fractures, 42, 45, 46, 64, 72, 79, 123, 124
Freedom of Information Act, 201
Funding, 10, 20, 31, 141, 202, 203, 262

G

Galactic cosmic rays, *see* Cosmic particles; Radiation exposure
Galvanic stimulation, 128
Gametes, 56, 58
Gastrointestinal system
see also Nutrition; Space motion sickness
absorption factors, 39
bowel infections, 111
bowel sounds, 52, 87
cancer, 103
ileus, 39, 123
medical events in Antarctica, 85, 86
medical events in space, 78, 81, 102-103; *see also "space motion sickness" infra*
medical events in submarines, 82
medications, 5, 52, 102
space motion sickness (SMS), 5, 39, 51-52, 72, 78, 87, 91, 102, 138, 140
Gender factors, 5, 11, 59-60, 77
see also Reproductive system
bone mineral loss, 43, 58, 59
breast cancer, 86-89
clinical research, 5, 60, 72
data collection and access, 13
extravehicular activity, suits, 60
group interactions, 156, 170
immune system, 59
orthostatic hypotension, 59
radiation effects, 5, 59, 88-89
sexual behavior, 138, 142, 149; *see also* Contraception and contraceptives
General living conditions, *see* Living conditions, general
Genetic factors
behavioral effects, 162-163
bone mineral density, 5, 43, 45, 47, 72
databases, 47
informatics, 213
musculoskeletal system, 5, 43, 45, 47
radiation exposure, 5, 56-57, 67, 72
reproductive system effects of radiation, 5, 56-57
risk assessment, general, 86-89
Goldin, Daniel, 189, 261-262
Gravity
see also Microgravity
artificial, 5, 72, 251
exposure to gravity following missions, 30, 48, 97-98, 129
Ground support personnel, 77, 148, 252
behavioral issues, 108-109, 140, 164, 166, 168
emergency and continuing care, 118
mission control, 108, 136, 140, 150-151
space crew interactions with, 108-109, 140, 152-158, 166, 168
Group interactions, 2, 21, 28, 142, 145-157, 160-161, 164-168, 170
analog environments, 145-146, 152, 156, 159-160, 162, 164, 165-167
cultural factors, 149, 156, 164-165, 169
disruptive, 14, 108, 154-156, 168, 170
gender factors, 156, 170
leadership, 10, 108, 147, 148
privacy while onboard spacecraft, 138, 142, 144

screening, selection, and training, 77, 146-147, 148, 149, 158-165, 169, 170
Growth, height, 26
Gynecologic health issues, 5, 103-105
contraception and contraceptives, 60, 104
menstruation effects, 5, 43, 58-59, 72, 103, 104
pelvic examinations, 88

H

Habitability, *see* Living conditions, general
Hazardous substances, 92, 93-94
anesthesia and pain management, 118, 119-120
clinical research, 65
countermeasures, 8, 94, 113
spacecraft maximum allowable concentrations (SMAC), 93-94, 143
Headaches, 110, 140
Health care personnel, 131-133, 255-256
anesthesia and pain management, 118
behavioral factors, 150
emergency and continuing care, 118
flight surgeon, 174, 175, 180-181, 215
organizational requirements, 193-194
training of, 30, 34, 214-216
Health registers, 83, 141
Hearing, *see* Auditory perception
Height, *see* Growth, height
Hematology, 61, 105-106
anemia, 39, 105, 106, 113
bone density markers, 43, 45, 46
clinical research, 9, 62, 64, 106, 113
erythropoietin, 40, 105
surgery, 121, 122, 125
High-energy particle radiation, *see* Radiation exposure
Historical perspectives, 3, 18, 26, 38, 137, 197
see also Analog environments
cardiovascular care, 97
clinical data collection, 6, 10
confidentiality, doctor-patient, 175-176
crew selection, 12
injury, responses to, 121
Mars exploration, 23
medical events, 78, 81
microgravity effects, 37
Hormonal contraceptive agents, 60, 104
Hormone replacement therapy, 9, 58-59, 104, 113
Hormones, *see* Endocrine function
Hubble Space Telescope, 202
Human Exploration of Mars: The Reference Mission of the NASA Mars Exploration Study Team, 24
Hydration, *see* Dehydration
Hydrostatic pressure, 39
Hypercalciuria, 44

I

Ileus, 39, 123
Immune system, 39, 153, 212
cell-mediated immunity, 105, 258
clinical research, 9, 62, 64, 66, 106, 113
gender factors, 59
injury, responses to, 121, 126
medical events in space, 81, 212
surgery, 121, 126
Infections, 127
abscesses, 82, 84, 127
analog environments, 82, 84, 85, 105, 126
bowel, 111
cell-mediated immunity, 105
clinical research, 9, 62, 64, 66
medical events in Antarctica, 85, 86
medical events in space, 9, 81, 105, 111
medications, 127
preflight isolation, 105
urinary, 111
wounds, 9, 105, 106, 113, 126
Informatics, 9, 24, 61, 208, 212-214, 217
Infrastructure, *see* Organizational factors

Injury, *see* Trauma
Institutional factors, *see* Organizational factors
Integration, 193-194, 197, 198, 218
  comprehensive health care system (NASA), 7, 9, 21, 40, 70-71, 190-191, 195-199 (passim)
  interagency coordination, 31, 73, 175, 193, 194, 195-196, 198, 199, 200, 203-208, 217
International cooperation, 198, 207, 217, 251
  behavioral health problems, 138, 140
  committee charge, 31
  ethical issues, 174, 186, 187
  risk assessment, 114
  screening, selection and training, 158-159
  standards, general, 8, 16, 17, 107, 219
International crews, 11, 24, 25, 149
  behavioral issues, 107, 137, 145, 170, 168
  ethical issues, 174, 186
International Space Station (ISS), 2, 3, 11, 17, 34, 38, 73, 77, 188, 190, 191, 202, 207, 212, 249, 251
  behavioral issues, 150, 152-153, 159
  bone mineral density, microgravity, 3, 40
  clinical research, 41, 73
  countermeasures research, 41-42, 68
  emergency and continuing care, 117, 190
  ethical issues, 181-186, 186, 187
  informatics, 213
  injury, responses to, 121
  launches from, 25
  mental health screening, 108
  radiation exposure, 96
  rehabilitation measures, 129-130, 190
  space motion sickness (SMS), 52
Intestines, *see* Gastrointestinal system
Isolation, 3, 10, 11, 13, 14, 25, 28, 29, 138, 140, 142, 168, 189, 258-259
  see also Communication and communications technology
  neurological effects, 53
  preflight, infection rates, 105

J

Johnson Space Center (JSC), 173, 183, 190, 192, 193, 195, 196, 248, 252

K

Kidneys, *see* Renal system

L

Language factors, 107, 140, 164-165, 169
  computer analysis of verbal communication, 154-155
Laproscopic surgery, 104, 124, 125
Laryngeal mask, 119
Leadership
  crew, 10, 108, 147, 148
  NASA organization, 20, 21, 192, 193, 194-195, 198-199
Legislation, 201-202
  see also Regulatory issues
  Freedom of Information Act, 201
  Privacy Act, 16, 175-176, 187, 201
Life on the Screen: Identity in the Age of the Internet, 109
Lighting in spacecraft, 5, 28, 142
  sleep and circadian rhythm, 53, 55, 144
Living conditions, general, 15, 65, 77, 134, 138, 142-143, 145, 161, 167
  see also Confined environments; Isolation; Lighting in spacecraft
  noise effects and controls, 8, 28, 93, 95
  paired transport vehicles, 149
  privacy onboard spacecraft, 138, 142, 144
Longitudinal studies, 15
  bone mineral loss, 45
  death, 78
  epidemiological data, 1, 9, 59, 197
  medical events in Antarctica, 83-86
  medical events in space, 78-79
  medical events in submarines, 82
  mental health issues, 109-110, 157-158

Longitudinal Study of Astronaut Health, 176, 209, 212, 249
Lucid, Shannon, 26
Lungs, *see* Respiratory system

M

Males, *see* Gender factors
Malnutrition, 65, 89
Mars missions, 11, 23-25, 28, 30, 38, 75, 137
  bone mineral density loss, 27, 44-45
  countermeasures, 68
  emergency and continuing care, 123-125, 132
  mental health issues, 107
  orthostatic hypotension, 49
  surgical procedures, 123-125
Medical devices, 19, 112, 129
  see also Robotics
  anesthesia and pain management, 118-119, 120
  electrical stimulation, 128, 129
  laryngeal mask, 119
  nanotechnology, 24, 61-62, 194
  surgery, 124-125
Medications
  see also Pain management and anesthesia
  bone mineral loss, 5, 44, 46, 47, 72
  clinical research, general, 5, 55, 204
  committee charge, 31
  databases, 8, 91, 113
  gastrointestinal system, 5, 52, 102
  hormonal contraceptive agents, 60, 104
  hormone replacement therapy, 9, 58-59, 104, 113
  infections, 127
  medical events in space, 79, 102
  mental illness treatment, 108
  muscle loss, 48, 72
  organizational requirements, 204, 217
  orthostatic hypotension, 49, 72
  pancreatitis prevention, 103
  pharmacodynamics, 59, 60, 65, 86, 90-91, 113
  pharmacokinetics, 5, 72, 90-91, 110, 113
  psychomotor functions, 55
  sleep, 53-54, 55
  space motion sickness, 5, 52
Melatonin, 54
Menstruation effects, 5, 43, 58-59, 72, 103, 104
Mental health and illness, 15, 28, 30, 64, 77, 106-110, 113, 138-139
  see also Behavioral health; Isolation; Mood effects; Screening, selection and training
  analog environment data, 80-81, 82, 85, 86
  cultural issues, 107, 149, 156, 158
  disruptive crew interactions, 14, 108, 154-155, 168, 170
  end-of-life care, 9, 130, 134
  intramission/postmission problems, 106, 107, 108-110, 153-154, 157-158
  longitudinal studies, 109-110, 157-158
  mission control, 108, 136, 140, 150-151
  NASA efforts, 106-107, 109-110
  suicidal and homicidal ideation, 108, 154
Metabolic processes
  see also Endocrine function
  bone mineral density, 43, 44-46, 47, 60-61, 87, 259
  injury, responses to, 121, 123
  medical events in space, 81
  nutrient absorption in microgravity, 89-90
  pharmacokinetics, 5, 72, 90-91, 110, 113
  renal stones, 61, 111-112
Microbiology
  see also Immune system; Infection
  anesthesia and pain management, 118, 119, 120
  contamination, 94-95, 113
  dental care, 99
Microgravity, 3, 6, 26, 29, 37-39, 40, 138, 189, 191, 249, 261, 262
  see also Bone mineral density

airway management during anaesthesia, 119
analog environments and, 76
artificial gravity, 5, 72, 251
body mass, 90, 127, 128
cardiovascular system, 8, 49-50, 97
clinical research, 5, 6, 9-10, 17, 42-43, 45, 46-47, 62-69
countermeasures, 40-42, 48, 56, 62-69, 72, 127-128; *see also "artificial gravity" supra*
decompression sickness, 59, 67, 92-93
emergency and continuing care, 121-123, 125, 127, 128, 129
ergonomic issues, 95-96, 113
exposure to gravity following, 30, 48, 97-98, 129
injury, responses to, 121, 122
menstruation effects, 5, 43, 58-59, 72, 103, 104
muscle loss, 9-10, 39, 47-48, 55-56, 62, 64, 66, 77, 127, 212
animal models, 48
countermeasures, 48, 56, 72, 127-128
neurological effects, 53, 56, 67
nutrient absorption, 89-90
orthostatic hypotension, 5, 39, 48-49, 50, 53, 59, 72, 97, 102, 110, 190
peripheral nervous system, 56
pharmacodynamics, 59, 60, 65, 86, 90-91
pharmacokinetics, 5, 72, 90-91, 110
physical examinations in, 87-88
radiation, synergistic effects, 65
reproductive system, 5, 8, 56
menstruation effects, 5, 43, 58-59, 72, 103, 104
space motion sickness (SMS), 5, 39, 51-52, 72, 78, 87, 91, 102, 138, 140
standards, 8, 113
surgery, 121-123, 125
urinary system, 39, 42, 44, 50, 60-61; *see also* Renal system
Mir space station, 249
behavioral problems, 140, 141, 153
bone mineral loss, 43, 44, 128
medication usage, 84
mental health issues, 106, 107
rehabilitation measures, 129
sleep and circadian rhythm, 53
surgical wounds, 122-123
Mission control, 108, 136, 140, 150-151
Model Trauma Care System Plan, 192
Mood effects, 8, 14, 86, 106, 113, 139-142, 155, 259
sleep and circadian rhythm, 53
Moon missions, 34, 38
Motion sickness, *see* Space motion sickness
Motor functions, *see* Psychomotor functions
Musculoskeletal system, 203
see also Bone mineral density
ataxia, 39, 48
body mass, 90, 127, 128
clinical research, 5, 9-10, 42-43, 45, 46-47, 62, 64, 66, 72
fractures, 42, 45, 46, 64, 72, 79, 123, 124
gender factors, 43, 58, 59
genetic factors, 5, 43, 45, 47
medical events in Antarctica, 85, 86
medical events in space, 79, 81
muscle loss, 9-10, 39, 47-48, 55-56, 62, 64, 66, 77, 127, 212
animal models, 48
countermeasures, 48, 56, 72, 127-128
pain management, 48
nutrition, 48
peripheral nervous system, 55-56
surgical wound healing, 122-123

N

Nanotechnology, 61, 194
sensors, 24, 61-62
National Aeronautics and Space Administration (NASA), 24, 189-191
see also Organizational factors

Ames Research Center, 193
behavioral issues, 138, 139, 141, 143, 146, 147, 148, 150-151, 154, 156-159, 160, 162, 164-165, 170
astronaut esprit as a problem, 174-175, 182-183
cardiovascular effects of microgravity, 50
committee charge and methodology, *xi-xiii*, 2-3, 29-33, 247-260
committee recommendations, 1, 6-21, 40, 69-73, 114, 135, 170, 187, 191, 202-219 (passim)
comprehensive health care system, 7, 9, 21, 40, 70-71, 190-191, 195, 196, 197, 198, 199, 200, 201, 208-209, 216, 218-219
countermeasures, general, 27, 68, 193, 194-195
behavioral, 156-157
bone mineral loss, 40-42, 46
Critical Path Roadmap project, 4, 62-69, 200, 211, 250
emergency and continuing care, 126, 129-130, 134-135, 190, 192, 210-212
ethical issues, 174, 175, 181-187
funding, 10, 20, 31, 141, 202, 203, 262
gamete preservation, 57
Johnson Space Center (JSC), 173, 183, 190, 192, 193, 195, 196, 248, 252
leadership, crews, 148
leadership, NASA organization, 20, 21, 192, 193, 194-195, 198-199
major health and medical events, 76-77
mental health of crew, 106-107, 109-110
neurological clinical program, 111
neurovestibular function, 53
physical training and rehabilitation, 129-130
physiological monitoring, 61-62
pulmonary system, 51
radiation exposure, 96
risk assessment, general, 3-4, 38
space medicine defined, 34
surgical procedures, 126
National Cancer Institute, 62
National Institute of Arthritis and Musculoskeletal and Skin Diseases, 203
National Institute on Aging, 203
National Institutes of Health, 194, 199, 200, 203, 204, 208
National Science Foundation, 195
National Space Biomedical Research Institute, 10, 143, 156-157, 193, 196, 202, 250
Nausea, *see* Space motion sickness
Nephrolithiasis, 60, 111-112
Nervous system, 5, 39, 49, 110-111
see also Pain management and anesthesia; Psychomotor functions; Sleep and circadian rhythm; Space motion sickness
analog environments, 53
animal models, 53
ataxia, 39
baroreceptors, 49, 50
behavioral factors, 162, 163
central nervous system, 53, 55, 65, 110, 111, 120
clinical research, 5, 10, 66, 72, 111, 113
closed-head injuries, 111
confined environments, 53
injury, responses to, 121, 123
isolation effects, 53
medical events in Antarctica, 85, 86
medical events in space, 81, 110-111, 212
microgravity, 53, 56, 67
peripheral nervous system, 55-56
space motion sickness (SMS), 5, 39, 51-52, 72, 78, 87, 91, 102, 138, 140
Neurovestibular function, 10, 53, 62, 65, 67, 77, 81, 110
Noise effects and controls, 8, 28, 93, 95
Nutrition, 11, 51, 88-90, 138, 142, 143, 251
bioregenerative systems, 89, 143, 145
bone mineral loss, 44

clinical research, 5, 62, 63, 65
injuries, recovery from, 123
malnutrition, 65, 89
medical events in space, 81
muscle protein, 48

O

Occupational health model, 1, 16-17, 92-97, 178-180, 187, 201, 203-204
ergonomics, 95-96, 113
Office of Naval Research, 194
Oocytes, 57, 58
Organizational factors, 2, 6-7, 10, 19-20, 29-30, 191-219, 262
see also International cooperation
authority, 20, 21, 194-195
behavioral health, 20, 193
comprehensive health care system (NASA), 7, 9, 21, 40, 70-71, 190-191, 195-199 (passim), 200, 201, 208-209, 216, 218-219
ethical issues, 174-175, 182, 187
group functioning, 147, 158, 168
health care personnel, 193-194
integration, 193-194
interagency coordination, 31, 73, 175, 193, 194, 195-196, 198, 199, 200, 203-208, 217
leadership, crew, 10
leadership, NASA, 20, 21, 194-195
neurological clinical program, 111
standards, 191, 192, 193, 195, 200
international cooperation, 8, 16, 17, 107, 219
systems development, 199-200
Orthostasis, 5, 39, 48-49, 50, 53, 59, 72, 97, 102, 110, 190
Osteoporosis, 45, 47, 64, 212

P

Pain management and anesthesia, 9, 118-122, 129, 134
backaches, 79
endotracheal intubation, 118, 119
headaches, 110, 140
medical devices, 118-119, 120
muscle loss and, 48
screening, selection and training, 118, 120
Pancreas, 103
Parathyroid hormone, 258, 259
Performance, *see* Behavioral health; Task performance
Peripheral nervous system, 55-56
Pharmaceuticals, *see* Medications
Pharmacodynamics, 59, 60, 65, 86, 90-91
analog environments, 91
databases, 8, 91, 113
Pharmacokinetics, 5, 72, 90-91, 110
databases, 8, 91, 113
Phenotypes, bone mineral loss, 43, 47, 72
Physiological monitoring, 39, 61-62, 68, 207
behavioral adaptation, 154, 155-156, 169
bone mineral density markers, 5, 43, 45, 46
clinical research, 5, 61, 62, 63, 68, 72
Pituitary gland, 57, 58
Polar tri-idothyronine (T3) syndrome, *see* Tri-iodothyronine (T3)
Poliakov, Valeri, 26
Post-flight support and reintegration, 106, 107, 108-110, 153-154, 157-158
Pregnancy, 60, 103
see also Contraception and contraceptives
Preventive measures, 28, 40, 99, 189-190, 210, 211, 213
see also Screening, selection, and training
surgical infections, 126
underreporting of problems, 175, 183, 187
Privacy, *see* Confidentiality and privacy
Privacy Act, 16, 175-176, 187, 201
Prostate gland, 112
Protocols
see also Clinical practice guidelines; Standards

confidentiality, 182-186
end-of-life considerations, 9, 130, 134
mental health, 108
organizational requirements, 200
physical exercise, 48
Psychomotor functions, 9, 39, 55, 64, 65, 134, 138, 212
computerized assessment, 155
medications, 55
recreational activities, 128-129
Public opinion/education, 6, 31, 114, 124, 131, 205-207
Pulmonary system, *see* Respiratory system

Q

Quality control, 8, 208, 217
see also Clinical practice guidelines; Ethical issues; Protocols; Regulatory issues; Standards
expert oversight, 8, 219

R

Race/ethnicity, *see* Ethnicity
Radiation exposure, 8, 27-28, 29, 62, 77, 92, 96, 113, 189, 249, 261
cancer and, 65, 88-89, 212
central nervous system, 65
clinical research, 28, 65
cosmic particles, 27, 96, 113
gender factors, general, 5, 59, 88-89
genetic factors, 5, 56-57, 67, 72
microgravity effects, synergies, 65
neurovestibular adaptation, 67
reproductive health, 5, 56-57, 67, 72
Real-time measurements, 5, 18, 134
bone mineral density, 46, 72
group interactions, 148
neurological effects, 52-53
surgical procedures, 123-124
Recovery systems, *see* Support and recovery systems
Recreation and leisure activities, 14, 128-129, 142, 144, 151
Regulatory issues, 1, 17, 182, 187
see also Legislation; Occupational health model; Standards
Rehabilitation practices, 7, 8, 11, 28, 67, 71, 73, 127-130, 190, 250
animal models, 127
International Space Station (ISS), 129-130, 190
Renal system
calculi, 39, 42, 44, 60-61, 66, 85, 111-112, 212
nephrolithiasis, 60, 111-112
vascular resistance, 50
Reproductive system, 56-60, 111-112
clinical research, 5, 57, 59, 72
exercise, 58
gametes, 56, 58
genetic factors, 5, 56-57
gynecological issues
hormones, 9, 57, 58, 72
replacement therapy, 9, 58-59, 104, 113
testosterone, 57
medical events in submarines, 82
microgravity, gynecological health issues, 5, 8, 43, 56, 58-59, 72, 103, 104
oocytes, 57, 58
pelvic examinations, 88
prostate gland, 112
radiation effects, 5, 56-57, 67, 72
sperm production, 56
Resources for the Optimal Care of the Injured Patient, 192
Respiratory system, 39, 51
airway management, 111, 118-119
endotracheal intubation, 118, 119
laryngeal masks, 119
medical events in Antarctica, 85, 86
medical events in space, 78, 81
medical events in submarines, 82, 84
nasal congestion, 78, 87
orthostatic hypotension, 5, 39, 48-49, 50
pulmonary function tests, 36
Reticulocytes, 105

Review of NASA's Biomedical Research Program, 38
Robotics, 262
Mars exploration, 24
surgery, 123-124

S

Sagan, Carl, 137
Screening, selection, and training, 28, 86, 158-165, 169, 170, 250, 256-257, 259
anesthesia and pain management, 118, 120
behavioral factors, 12, 77, 86, 107, 108
group interactions, 146-147, 148, 149, 158-165, 169, 170
underreporting of problems, 174-175, 183, 187
bone mineral density, 45, 63
cardiovascular health, 97, 98, 113
committee charge and recommendations, 8, 12, 13, 15, 29, 32, 170
computerized assessment, 154-155
crew diversity, 77
Critical Path Roadmap project, 63
emergency and continuing care, 123-125, 127, 131-132
ethical issues, 182-183, 187
gastrointestinal system, 102
group interactions, 77, 146-147, 148, 149, 158-165, 169, 170
health care personnel, 30, 34, 214-216
international cooperation, 158-159
Mars exploration, 25
mental health issues, 107, 108
organizational requirements, 189, 210, 214-216
paired transport vehicles, 149
prostate cancer, 112
rehabilitation measures, 127, 134
surgical skills, 123-125
Self-assessment, 108, 109, 139-140, 145, 164, 169
Self-monitoring, 145
Sex differences, *see* Gender factors
Sexual behavior, 138, 142, 149
see also Contraception and contraceptives
Shaw, Chuck, 75
Shuttle program, *see* Space shuttles
Skeletal system, *see* Musculoskeletal system
Skin diseases, 78-79, 81, 82, 84, 85, 86, 203
cell-mediated immunity, 105
preflight isolation, 105
Skylab, 27, 74
Sleep and circadian rhythm, 5, 39, 53-55, 64, 72, 78, 139, 142, 143-144
lighting in spacecraft, 53, 55, 144
medications, 53-54, 55
Small Clinical Trials: Issues and Challenges, 70, 207
Small n problem, 214-215
Social issues, 2, 21, 28, 64
see also Behavioral factors; Cultural factors; Group interactions; Public opinion
analog environment data, 80-81
gender factors, 59
group interactions, 2, 21, 28, 138-142, 145-149, 156
Socrates, 23
Solar particle events (SPEs), 27, 96, 113
Space Behavioral Assessment Tool, 14, 155, 168
Space Flight Cognitive Assessment Tool, 14, 155
Space Flight Fatigue Assessment Tool, 155
Space motion sickness (SMS), 5, 39, 51-52, 72, 78, 87, 91, 102, 138, 140
medications, 5, 52
Space shuttle program, 36, 77, 102, 116, 136, 190, 207, 248, 252
behavioral issues, 140, 155
bone mineral loss, 43
dehydration, 112
sleep and circadian rhythms, 144
toxic chemical exposure, 93

Space suits, 93
decompression sickness, 92-93
females, 60
Spacecraft maximum allowable concentrations (SMAC), 93-94, 143
Sperm production, 56
Standards, 9
see also Ethical issues; Protocols
cardiovascular health screening, 97
clinical care, 7, 97, 113, 210-212; *see also* Clinical practice guidelines
ergonomic, 95-96, 113
group interaction monitoring, 156
health status measures database, 133
international cooperation, 8, 16, 17, 107, 219
mental health, 107
microgravity effects, 8, 113
occupational health model, 1, 16-17, 92-97, 178-180, 187, 201, 203-204
organizational requirements, 191, 192, 193, 195, 200
international cooperation, 8, 16, 17, 107, 219
toxic substance concentrations (SMAC), 93-94, 143
A Strategy for Research in Space Biology and Medicine in the New Century, 38, 139
Submarines, 76, 80-83, 84, 85, 106, 140-141, 152, 162, 205, 256-257, 259
Suicide, 108, 154
Suits, *see* Space suits
Support and recovery systems, 14, 26, 138, 147, 149-158, 161, 164, 168-169
see also Ground support personnel
intramission/postmission problems, 106, 107, 108-110, 153-154, 157-158
mental health, 106, 107, 108-110, 153-154, 157-158
mission control, 108, 136, 140, 150-151
psychomotor functions, 55
Surgery, 28, 30, 80, 102, 103, 121, 122, 123-127, 136, 215
see also Wounds
analog environments, 126-127
anesthesia and pain management, 118-119, 123
animal models, 121, 123
appendicitis, 29, 82, 85, 88, 102, 104, 127
cholecystectomy, 102, 103
communications technology and, 123-124
ethical issues, flight surgeon, 174, 175, 180-181, 215
gynecological, 103
hematological considerations, 121, 122, 125
immune system effects, 121, 126
laproscopic, 104, 124, 125
medical devices, 124-125; *see also* Robotics
medical events in submarines, 83-86
microgravity during, 121-123, 125
musculoskeletal system, 122-123
robotic, 123-124
Systemic Multiple Level Observation of Groups, 156
Systems development
bioregenerative systems, 89, 143
complex adaptive systems, 148
organizational, 199-200
triage, 118, 134, 211-212

T

Task performance, 62, 64, 66, 68, 198, 249, 251
behavioral issues and, 137-172
gender factors, 60
mental health factors, 106-107
neurovestibular adaptation, 67
psychomotor functions, 9, 39, 55
sleep and circadian rhythm, 53
Telecommunications, *see* Communication and communications technology

Temperature factors, 5, 129
Mars, 23
sleep and, 55
Testosterone, 57
Titov, Vladimir, 26
To Err Is Human, Building a Safer Health System, 217
Toxic chemicals, *see* Hazardous substances
Training requirements, *see* Screening, selection, and training
Trauma, 29, 63, 65, 78, 79-80, 81, 121-123, 127
see also Surgery
anesthesia and pain management, 118-119, 123
animal models, 121
cardiovascular response, 121-122
clinical research, 63, 65
closed-head and spinal cord injuries, 111
emergency and continuing care, 118
fractures, 42, 45, 46, 72, 79
immune system, 121, 126
medical events in Antarctica, 85, 86
medical events in submarines, 82, 84, 85
microgravity, response effects, 121, 122
multiple casualties, 118, 134
nutrition, recovery from, 123
organizational factors, 192
wounds
Triage, 118, 134, 211-212
Tri-iodothyronine (T3), 101

U

Urinary system, 60-61
see also Renal system
bone mineral density markers, 5, 43, 45, 46, 60
clinical research, 5
countermeasures, 60-61, 112
infections, 111
medical events in space, 81, 111-112
medical events in submarines, 82
microgravity, 39, 42, 44, 50, 60-61
urinalysis, other than density markers, 61

V

Vestibular function, *see* Neurovestibular function
Visual perception, 53
Vomiting, *see* Space motion sickness

W

Weight, *see* Body mass
Weightlessness, *see* Microgravity
Windows Space Flight Cognitive Assessment, 14, 155, 168
Women, *see* Gender factors
Wounds, 42, 84, 105, 113, 122-123
see also Surgery
analog environments, 126, 127
animal models, 121, 123
clinical research, 9, 106
infections, 9, 105, 106, 113, 126
Wyle Laboratories, 195-196